Paying for University Research Facilities and Administration

Charles A. Goldman, T. Williams

with

David M. Adamson, Kathy Rosenblatt

Supported by the
Office of Science and Technology Policy

Science and Technology Policy Institute

RAND

The research described in this report was conducted by RAND's Critical Technologies Institute, now the Science and Technology Policy Institute, under Contract ENG-9812731.

ISBN: 0-8330-2805-7

Published 2000 by RAND
1700 Main Street, P.O. Box 2138, Santa Monica, CA 90407-2138
1333 H St., N.W., Washington, D.C. 20005-4707
RAND URL: http://www.rand.org/
To order RAND documents or to obtain additional information, contact
Distribution Services: Telephone: (310) 451-7002; Fax: (310) 451-6915;
Internet: order@rand.org

PREFACE

In 1998, Congress directed the White House Office of Science and Technology Policy (OSTP) to conduct an analysis of issues related to the ways universities recover the facilities and administrative costs (also known as indirect costs) they incur when performing research under federal grants and contracts. At OSTP's request, the RAND Science and Technology Policy Institute prepared this report to present objective information on facilities and administrative costs in U.S. higher education. This report addresses the issues raised by Congress, although its scope is limited to presenting factual information and analysis of alternatives. It does not take positions on policy options. The report should be of interest to scientists, higher education administrators, and federal, state, and local governments.

THE SCIENCE AND TECHNOLOGY POLICY INSTITUTE

Originally created by Congress in 1991 as the Critical Technologies Institute and renamed in 1998, the Science and Technology Policy Institute is a federally funded research and development center sponsored by the National Science Foundation and managed by RAND. The institute's mission is to help improve public policy by conducting objective, independent research and analysis on policy issues that involve science and technology. To this end, the institute

- supports the Office of Science and Technology Policy and other Executive Branch agencies, offices, and councils
- helps science and technology decisionmakers understand the likely consequences of their decisions and choose among alternative policies
- helps improve understanding in both the public and private sectors of the ways in which science and technology can better serve national objectives.

Science and Technology Policy Institute research focuses on problems of science and technology policy that involve multiple agencies. In carrying out its mission, the institute consults broadly with representatives from private industry, institutions of higher education, and other nonprofit institutions.

Inquiries regarding the Science and Technology Policy Institute or this document may be directed to:

Bruce W. Don
Director, Science and Technology Policy Institute
1333 H Street, N.W.
Washington, D.C. 20005
Phone: 202-296-5000, ext. 5351
Web: http://www.rand.org/centers/stpi
Email: stpi@rand.org

CONTENTS

FIGURES

TABLES

SUMMARY

Federal spending for scientific research at U.S. academic institutions amounts to approximately $15 billion each year, funding a variety of projects that improve human health, our understanding of the natural world, education, national defense, and other areas. About three-quarters of this amount supports the direct costs of conducting research, such as the materials and labor used to perform each project. The other one-quarter covers facilities and administrative (F&A) costs. F&A costs (sometimes called indirect or overhead costs) encompass spending on such items as facilities maintenance and renewal, heating and cooling, libraries, and the salaries of departmental and central office staff.

Higher education institutions are eligible for reimbursement of F&A costs related to federal grants and contracts. They do not necessarily receive full reimbursement for these costs, however. F&A reimbursement rates are set by negotiation between the federal government and each university, based on accounting data. There are also statutory limitations that apply to certain programs. In these cases, universities recover less than their negotiated F&A rate.

Congress has long taken an interest in facilities and administrative costs in higher education. In 1998, Congress asked for an investigation of issues related to this topic. At the request of the White House Office of Science and Technology Policy, we undertook analyses of these issues.

In conducting our analysis, we have been hampered, in some cases, because the government does not maintain convenient databases from which to extract the requested information. The accessible government data contain information on negotiated facilities and administrative rates. Our analysis of these data shows that these negotiated rates have remained about constant for a decade, but we lack data on actual federal outlays for F&A costs. The data we do have are consistent with the findings based on negotiated rates.

Because we have to rely on incomplete data for actual outlays by agencies and receipts by universities, we can only make approximations in these areas. On average, about 31 percent of total true costs appear to be for facilities and administration. The share of federal outlays that pays for F&A costs is somewhere between 24 and 28 percent, indicating that universities are sharing significantly in the facilities and administration costs. There are requirements in law for universities to share certain costs. In addition, universities voluntarily agree to share costs on federal projects. Overall, we estimate that universities are providing between $0.7 and $1.5 billion in facilities and administration costs that would be eligible for reimbursement based on their negotiated F&A rates. We estimate that universities are recovering between 70 and 90 percent of the facilities and administrative expenses associated with federal projects.

Because universities report a total level of support for research from their own funds of about $5 billion, it appears that these facilities and administration costs represent about one-fifth of the university funds devoted to research. A further portion of the $5 billion in university funds represents universities' sharing in the direct costs of some projects, in particular by subsidizing faculty time.

The universities are voluntary participants in this system. They offer and provide these funds to share the costs of research because they perceive good reasons to do so. Federal projects bring prestige to faculty in their careers and universities as institutions.

To analyze F&A rates further, we divide them into two major components. The administrative component includes salaries and expenses for accounting, general administration, sponsored projects administration, and departmental administration. Negotiated rates are subject to certain limits, including a cap on the amount of administrative costs that may be included. The facilities component includes expenses for the construction, operation, maintenance, and renovation of both buildings and equipment. Although there are no fixed caps on facilities costs, limitations and reviews apply.

As administrative rates have declined because of the imposition of the administrative cap, facilities rates have increased, leaving overall negotiated rates about constant since the late 1980s.

In terms of the reasonableness of F&A costs in universities, our direct evidence is limited. What evidence we have indicates that the underlying cost structures in universities have lower F&A costs than federal laboratories and industrial research laboratories. Because of specific limitations on university F&A reimbursement, such as the administrative cap, the actual amount awarded to universities for F&A costs is likely to be even lower than what cost structure comparisons would indicate.

Although universities clearly exercise some discretion in deciding how to staff administrative offices and how to construct facilities, many of the costs of facilities and administration derive from requirements in federal, state, and local law. These laws and regulations support a number of objectives, including the desire to protect the health and safety of humans and animals and to promote good stewardship for federal research funding. But they impose real costs.

Facilities rates have increased partly because of a change in federal policy that allows the inclusion of interest costs on new construction in rate negotiations. Universities appear to have undertaken modernization especially during the 1990s, increasing research space by 28 percent and resulting in increased costs for construction. The operations and maintenance component of rates has declined, perhaps because newer facilities are more efficient to operate.

If the federal government were to significantly reduce payments for facilities and administrative costs, universities might pursue various options to make up some of the difference. We do not know how universities would finance additional cost-sharing. The $5 billion in university funding for research already includes F&A costs on federal projects that the federal government does not reimburse. Universities faced with reduced federal reimbursement for facilities and administration might follow several strategies. They could reduce other projects within the $5 billion to provide more of that amount as cost-sharing for F&A costs. As an alternative, universities could slow investments in building new facilities or renovating old ones. Other possible sources of funds for greater cost-sharing on research could come from reducing internal funding for other missions, such as education, public service, or patient care. We lack data to indicate the choices that universities would make. It seems worthwhile to further investigate the options for universities to shift funding and the consequences of those shifts before contemplating major changes in reimbursement of F&A costs.

One alternative to direct federal reductions in payments for F&A costs is to examine the laws and regulations that give rise to costly requirements on university facilities and administration. If some of these requirements could be streamlined, universities could reduce costs and the federal government could lower payments for F&A costs without forcing universities to shift resources from other programs.

Overall, the research partnership between the federal government and universities is widely praised for its contributions to the public welfare. Facilities and administrative costs are real to both the government and universities. These costs, like all research costs, are shared among the federal government, state governments, universities, industry, and private donors. The exact amounts shared by each participant in the system are subject to policy debate and negotiation. This report provides up-to-date quantitative and qualitative data on facilities and administrative costs to inform that policy debate.

ACKNOWLEDGMENTS

The authors thank the Department of Defense Office of Naval Research and the Department of Health and Human Services Division of Cost Allocation. Both agencies cooperated fully with the RAND research team to make data available, discuss policies and procedures, and review several drafts of this report. We are also grateful for assistance from the National Institutes of Health, the Department of Agriculture, the Office of Management and Budget, and the Council on Governmental Relations.

Sybil Francis and Arthur Bienenstock of the White House Office of Science and Technology Policy provided timely and useful guidance throughout the project.

Several reviewers provided thoughtful comments that strengthened this report.

GLOSSARY

AHM	Acutely hazardous materials
AWA	Animal Welfare Act
CAS	Cost Accounting Standards
CERCLA	Comprehensive Environmental Response, Compensation, and Liability Act (Superfund)
Circular A-21	Office of Management and Budget Circular, *Cost Principles for Educational Institutions*
Circular A-110	Office of Management and Budget Circular, *Uniform Administrative Requirements for Grants and Other Agreements with Institutions of Higher Education, Hospitals, and Other Non-profit Organizations*
Circular A-133	Office of Management and Budget Circular, *Audits of States, Local Governments, and Non-profit Organizations*
COGR	Council on Governmental Relations
DHHS	Department of Health and Human Services
DHHS DCA	Department of Health and Human Services Division of Cost Allocation
Direct costs	Costs that can be closely related to a specific project, such as salaries and materials
DS-2	Disclosure Statement, Form DS-2
DoD	Department of Defense
DoE	Department of Energy
EPA	Environmental Protection Agency
F&A	Facilities and administration
F&A costs	Facilities and administrative costs (term used in higher education for indirect costs as defined below)

FDA	Food and Drug Administration
FOIA	Freedom of Information Act
HMTA	Hazardous Materials Transportation Act
IACUC	Institutional Animal Care and Use Committee
Indirect costs	Costs that are shared among multiple projects, such as construction, operation, maintenance, and administration
MTDC	Modified total direct costs (direct costs less exclusions, such as equipment and large subcontracts)
NASA	National Aeronautics and Space Administration
NIH	National Institutes of Health
NSF	National Science Foundation
OMB	Office of Management and Budget
ONR	Department of Defense Office of Naval Research
OSHA	Occupational Safety and Health Administration
OSTP	White House Office of Science and Technology Policy
PHS	Public Health Service of DHHS
R&D	Research and development
RCRA	Resource Conservation and Recovery Act
RMPP	Risk Management and Prevention Program
SBIR	Small Business Innovation Research (Program)
STTR	Small Business Technology Transfer (Program)
TSCA	Toxic Substances Control Act
UCA	Utility Cost Adjustment
USDA	Department of Agriculture

Chapter One

INTRODUCTION

Federal spending for scientific research at U.S. academic institutions amounted to $15.1 billion in 1997. As Figure 1.1 shows, the federal government is the largest source of funding for research in colleges and universities. Other external sources provide substantial funds as well: about $2 billion each from industry, state and local governments, and a combination of other funders, mostly foundations and private gifts. After the federal government, the largest supporter of university research is the universities themselves from their own funds. Each year universities direct resources they control to support about $5 billion in research.

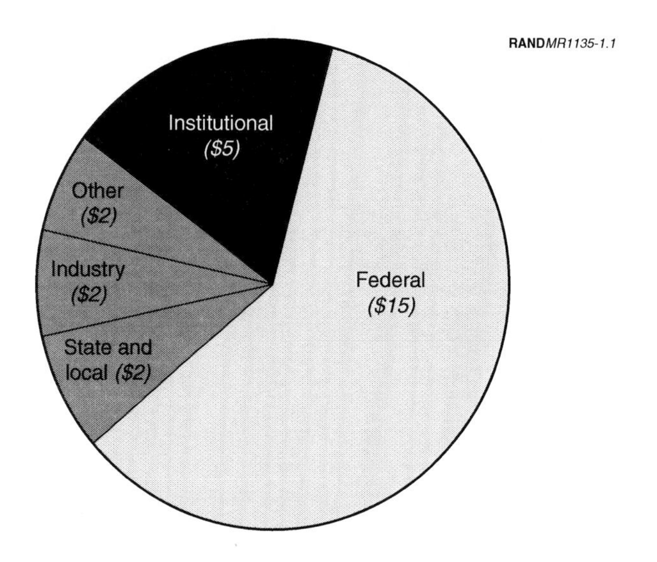

SOURCE: R&D Expenditures, FY 1997. NSF WebCASPAR system.
NOTE: Values as reported by colleges and universities.

Figure 1.1—Funding for Research in Higher Education, 1997 (Billions)

Looking at federal support in more detail, we see six agencies that sponsor most of the research in colleges and universities. As shown in Figure 1.2, one agency, the Department of Health and Human Services (DHHS), accounts for more than half of the total federal outlay. The DHHS includes the National Institutes of Health (NIH), which organizes almost all of this agency's academic research funding. Five other agencies—the National Science Foundation (NSF), the Department of Defense (DoD), the Department of Agriculture (USDA), the National Aeronautics and Space Administration (NASA), and the Department of Energy (DoE)—account for almost all the rest of federal research funding for colleges and universities.

Just as research funding is concentrated in a few agencies, most of the funds go to a relatively small number of institutions. There are more than 4,000 accredited institutions of higher education in the United States. Of these, about 460

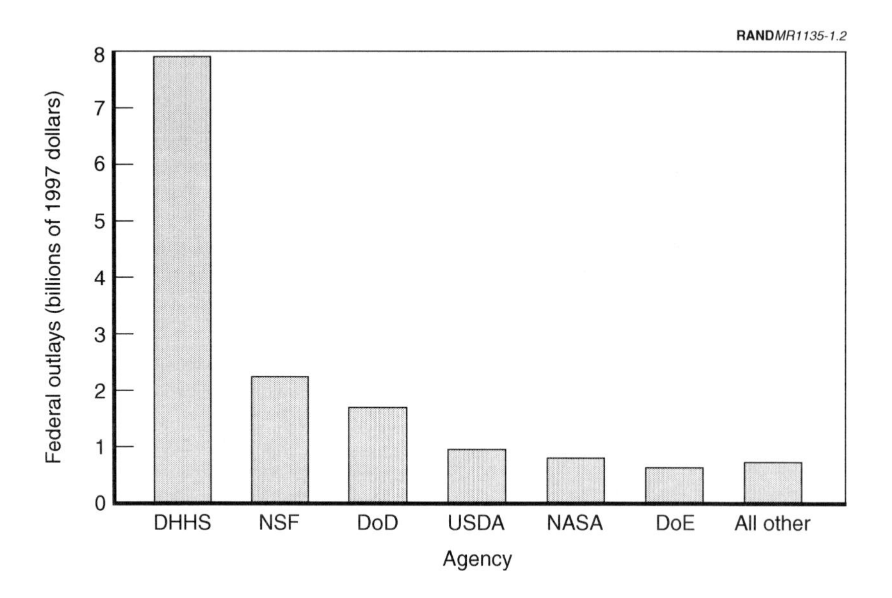

SOURCE: Federal Obligations for Sciences and Engineering, FY 1997. NSF Web-CASPAR system.

NOTE: Figures represent "Research and Development" as defined by the NSF. For the Department of Agriculture, this includes the federal funding for agricultural extension programs as well as research project grants.

Figure 1.2—Federal Outlays for Research in Higher Education by Agency, 1997

report receiving some federal research funding on an NSF survey.[1] Major recipients are a smaller set. The top 50 recipients of federal research support account for 60 percent of total spending. The top 150 recipients account for more than 90 percent of the total.[2]

The partnership in research between the federal government and U.S. research universities has been beneficial to both. The partnership has been widely praised for advancing scientific knowledge, improving the quality of life of Americans, contributing to the nation's prosperity, strengthening its national security, promoting technological innovation, and training the future scientific workforce that will continue these advances in the future. Recent congressional calls for doubling science budgets across the board indicate the high regard policymakers have for this partnership.

In this partnership, the partners not only share some important objectives but also experience some divergence in their interests. Federal agencies naturally try to stretch their budgets and seek an equitable cost to the taxpayer. Universities want projects to have adequate resources to cover their costs. If those resources are not provided—or are not fully provided—by federal agencies, then universities must cover the remaining costs from other sources. As shown in Figure 1.1, universities do cover a substantial amount of the total research budget.

Universities combine many activities and sources of funds. These combinations can make it ambiguous exactly who is paying for what. The costs of maintaining buildings, for example, may be properly shared among instruction, research, and public service functions. Because universities pursue many objectives simultaneously (including various teaching programs and research projects), they incur significant shared costs that benefit multiple objectives. The federal government has developed a set of procedures for allocating these costs to the multiple objectives, which we will describe in Chapter Two.

PERSPECTIVES ON FACILITIES AND ADMINISTRATIVE COSTS

Based on the analysis in this report, about three-quarters of federal outlays support the direct costs of conducting research, such as the materials and labor

[1]The NSF survey is the Survey of Federal Science and Engineering Support to Universities, Colleges, and Nonprofit Institutions, 1997. As noted in Chapter Two, the cognizant agencies have current F&A rate agreements with about 1,000 higher education institutions. The number is higher than the survey figure for several reasons. Some of these may have rate agreements but not have actively received funds in a given year covered by the survey. In addition, some university units, such as medical schools, may be separated into independent institutions for rate purposes, increasing the number of rate agreements.

[2]Source: Federally Financed R&D Expenditures for 1997. NSF WebCASPAR system.

used to perform each project. The other one-quarter covers facilities and administration (F&A) costs. F&A costs (sometimes called indirect or overhead costs) encompass spending on such items as facilities maintenance and renewal, heating and cooling, libraries, and the salaries of departmental and central office staff.

Allocating shared costs to projects is a simple concept, but the detailed rules governing how to do the allocation for research universities are complex. The rules for recovering facilities and administrative costs have evolved through about 15 revisions since they were first standardized in 1958 by the Office of Management and Budget (OMB). Some changes were initiated by the Executive Branch, and some were developed in response to congressional concerns, discussed in more detail in the following paragraphs. Many of these changes were intended to prescribe standard ways of accounting for costs and seeking reimbursement. In addition to OMB rules, codified in OMB Circular A-21, colleges and universities face a number of other requirements imposed by specific agencies and funding mechanisms governing which of their F&A costs are eligible for reimbursement. Higher education institutions may also voluntarily share the costs of facilities and administration. As a result of mandatory and voluntary cost-sharing, federal outlays for F&A costs amount to less than the full documented costs on campus. In this report, we estimate that the federal government reimburses universities somewhere between $3.6 and $4.2 billion per year for F&A costs. The federal government does not reimburse an additional $0.7 to $1.5 billion in F&A costs allocated to federal projects. Universities also share in the direct costs of projects, for example, contributions of faculty time. That form of cost-sharing is not included in these calculations.

As noted above, cost-sharing in general makes it difficult to be specific about who is actually paying for what. The partnership is characterized by various features. A large number of universities compete for federal research grants and contracts. A small number of agencies provide funds. Universities have information about their cost structure that may be difficult for agencies to verify. Universities must make long-term investments in people and facilities in anticipation of their ability to recover costs from federal projects. Universities must construct research facilities with a lifetime of several decades and bear the risk that their fortunes in federal funding may change or that the rules of the cost recovery framework may change over that time. In a very similar way, universities grant tenure to scientists. Tenure is provided as an incentive to encourage independent thinking, which is especially important in scientific research. But tenure implies a career-long commitment to a scientist. The university bears the risk of these investments in people and facilities. This system promotes healthy competition among institutions and researchers, but it requires the universities to bear the risks of their investments.

The codification of rules in OMB Circular A-21 in part limits the risk borne by the universities. In general, an environment in which the rules are subject to frequent revisions is more risky. When a university board of trustees faces the decision of whether to invest in building a new research facility, it considers how likely the university is to recover costs from the federal government. To the extent that cost-recovery rules are stable over time, the prospects for cost recovery are more certain. Although we do not have hard evidence on how changes in rules affect decisionmaking, private conversations with university board members indicate that they consider the stability of the federal cost-recovery system in their building decisions. Starting in 1981, the government allowed universities to seek reimbursement not only for depreciation of buildings but also for interest costs for construction as described in Chapter Five. This provision reduced the risks of investments in buildings. Although the evidence is not conclusive, a substantial increase in the quantity and quality of research facilities occurred after this provision went into effect.

We observed differences in agency policies and practices for cost recovery. Some agencies stay close to full cost recovery for universities; others reimburse significantly less than full project cost. If the government as a whole significantly underreimburses university costs, then universities will seek ways to make up the difference in their personnel and facilities costs from other sources.

Universities, according to the data in Figure 1.1, already cover much of the costs of research from funds they control. Clearly, universities value federal research support and are willing to accept somewhat less than full cost recovery. They are, after all, already sharing in the costs of the overall research enterprise. But there are limits to how much a given university can share in research costs before other programs must give way. A university must also provide education to its students and perhaps other functions, such as patient care or agricultural extension service. If federal support for research is reduced, whether for salaries or for facilities, universities may have to cut back in these other areas. If universities do not cut back in other areas, they may avoid constructing new facilities, renovating existing facilities, or investing in the careers of scientists. Without high-quality facilities and personnel, universities may shift their research focus or even reduce their overall research activity. Nonetheless, universities do support a good deal of research, including part of the facilities and administrative costs for federal projects.

Some research programs align closely with the interests of other university funders, including state governments, private donors, and students. These types of programs make it easier for universities to share costs because the objectives of more than one funder are simultaneously satisfied. When a federal agency supports research that is not closely aligned with the interests of these other funders, the federal agency should expect to pay more of the full cost of research.

CONGRESSIONAL INTEREST IN FACILITIES AND ADMINISTRATIVE COSTS

As the research partnership between the federal government and universities developed, federal agencies developed principles for reimbursing both the direct costs of research and some of the costs of facilities and administration. The reimbursement of these costs has long been the subject of congressional interest. In the late 1980s, there were some widely publicized incidents of alleged overcharges for F&A expenses, and in a few cases, universities returned some federal funds.

In 1991, the House Science Committee, working through the vehicle of the National Science Foundation Authorization Act, expressed its intention that both administrative and facilities costs should be restricted. For administrative costs, the committee intended that a specific numerical cap apply to recovery (26 percent of modified total direct costs as further explained in Chapters Two, Four, and Five). For facilities, the committee intended a requirement that whatever reimbursement was received by a university the full amount must be applied to research buildings and equipment and to no other purpose (House Science Committee, 1991). Although the NSF Authorization Act was not passed in that session, OMB did modify Circular A-21 to incorporate both of these provisions governmentwide. The provisions adopted can be seen in the history of changes to OMB Circular A-21, as detailed in Appendix A.

The language of the 1991 NSF Authorization Act also called for study to define more carefully the cost categories used in facilities and administrative rates. Again, although the Act was not passed by Congress, OMB did study and define cost categories, issuing a new version of Circular A-21 in 1993, which incorporated these defined categories and a number of other changes.

Subsequent sessions of Congress continued to express concern over the level of F&A costs. In 1995 and 1997, the House Science Committee again took up an NSF Authorization Act. During discussions of this act, the committee advocated a shift in how research funds were allocated.

> The Committee continues to be concerned that too great a share of academic research funds may be allocated to indirect costs. According to the President's budget, over one-quarter of the $12 billion the government spends on research at universities and colleges are used to cover indirect costs. While the government has a responsibility to reimburse that portion of the overhead directly associated with carrying out federally sponsored research, the Committee is

> concerned that the current system of indirect cost payments is consuming too large a share of a limited research budget. (House Science Committee, 1997.)

The committee was not seeking to reduce funding for research. The committee desired to maintain the same overall level of funding for universities but sought to shift the balance more toward direct costs and away from facilities and administrative (indirect) costs.

> The Committee believes that any resultant savings in indirect cost payments should be used to increase overall federal research support. (House Science Committee, 1997.)

Specifically, the committee called on the Executive Branch to propose methods to reduce outlays on facilities and administrative costs by 10 percent. The 1997 version of the Act was passed by the House and sent to the Senate for consideration.

In the Senate, the Committee on Labor and Human Resources echoed the concerns of the House Science Committee.

> The committee is greatly concerned about the rising cost of the administration and delivery of scientific research and higher education. (Senate Committee, 1997.)

The Senate committee connected concerns about the cost of research to state and federal regulations as well as possible influence on tuition rates for college students.

> In recent years university administrators have cited State and Federal regulatory burdens as well as the unreimbursed costs of conducting scientific research as contributors to the rapid growth in the cost of attending college. (Senate Committee, 1997.)

The Senate committee did not preserve the House's desire for a study of how to achieve a 10 percent reduction in facilities and administrative costs. Instead the Senate substituted a request to study specific concerns related to the federal government's role in reimbursing these costs. One concern was how the federal government fared in comparison with other research sponsors.

> In 1992, the Department of Health and Human Services inspector general testified that many schools charge the Federal Government higher indirect cost rates than they charge other research sponsors, including "foundations, public corporations, and foreign Governments. . . . It appears clear that schools may be looking to the Federal Government to cover the overhead associated with research performed for non-Federal and foreign entities." (Senate Committee, 1997.)

The Senate version of the NSF Authorization Act was passed May 12, 1998, including a request to the White House Office of Science and Technology Policy (OSTP) for a detailed report on six issues related to facilities and administrative costs. The six issues are quoted in the following section. This version of the Act was subsequently passed by the House and signed into law by the President on July 29, 1998.

Some observers believe that F&A spending consumes an increasing share of federal research dollars, with a corresponding decrease in funds going directly to researchers. *The data presented in this report do not support this view.* Overall, the system appears stable. According to the available data, F&A spending as a percentage of project cost has remained about level for at least a decade. In addition, F&A spending at colleges and universities is generally slightly lower than at other types of research institutions, such as federal laboratories and industrial research laboratories.

PURPOSE AND ORGANIZATION OF THIS REPORT

As explained above, in the NSF Authorization Act of 1998, Congress directed OSTP to conduct an analysis of six issues. At the request of OSTP, the RAND Science and Technology Policy Institute compiled and analyzed current information to assist OSTP, Congress, and the public to understand and discuss policy choices for indirect cost recovery. The analysis was structured around the six issues raised by Congress:

> **Issue 1:** analyze the federal indirect cost reimbursement rates (as the term is defined in Office of Management and Budget Circular A-21) paid to universities in comparison with federal indirect cost reimbursement rates paid to other entities, such as industry, government laboratories, research hospitals, and nonprofit institutions.
>
> **Issue 2:** analyze the distribution of the federal indirect cost reimbursement rates by category (such as administration, facilities, utilities, and libraries) and by the type of entity; and determine what factors, including the type of research, influence the distribution.
>
> **Issue 3:** analyze the impact, if any, that changes in Office of Management and Budget Circular A-21 have had on
>
> > the federal indirect cost reimbursement rates, the rate of change of the federal indirect cost reimbursement rates, the distribution by category of the federal indirect cost reimbursement rates, and the distribution by type of entity of the federal indirect cost reimbursement rates; and

the federal indirect cost reimbursement (as calculated in accordance with Office of Management and Budget Circular A-21), the rate of change of the federal indirect cost reimbursement, the distribution by category of the federal indirect cost reimbursement, and the distribution by type of entity of the federal indirect cost reimbursement.

Issue 4: analyze the impact, if any, of federal and state law on the federal indirect cost reimbursement rates.

Issue 5: analyze options to reduce or control the rate of growth of the federal indirect cost reimbursement rates, including such options as benchmarking of facilities and equipment cost, elimination of cost studies, and mandated percentage reductions in the federal indirect cost reimbursement, and assess the benefits and burdens of the options to the federal government, research institutions, and researchers.

Issue 6: analyze options for creating a database that would serve two functions: tracking the federal indirect cost reimbursement rates and the federal indirect cost reimbursement and supporting analysis of the impact that changes in policies with respect to federal indirect cost reimbursement will have on the federal government, researchers, and research institutions.[3]

This report presents the results of RAND's analysis. Issues 1 through 4 above are essentially factual investigations. RAND has compiled and analyzed available data in support of these issues. For Issues 5 and 6, RAND, in its role as an objective analyst, can present options for both OSTP and Congress to consider. RAND does not take a position on the various alternatives presented; that is the purview of the policymaking community.

This report continues with a background discussion on the principles, procedures, and methods for determining rates of reimbursement of facilities and administrative costs. Following the background discussion, six chapters correspond to each of the six issues identified by Congress. The main body of the report ends with a brief conclusion. Two Appendixes contain additional information on OMB Circular A-21: a detailed history of changes and a description of rate types.

---

[3]Section 203 of NSF Authorization Act of 1998, Public Law 105-207.

Chapter Two

BACKGROUND: HOW UNIVERSITIES RECOVER F&A COSTS FROM THE FEDERAL GOVERNMENT

This chapter begins by explaining the general principles under which universities recover facilities and administrative costs, the formal procedures for calculating F&A costs and negotiating F&A cost-recovery rates, and the methods for calculating how much of federal R&D outlays go to F&A costs.

There are limits set by statute and agency policy on universities' ability to recover costs from the federal government for F&A spending on government-funded research. By contrast, commercial firms that do business with the federal government generally can recover the full costs of government-related business expenses. Historically, the underlying reasons for limiting universities' cost recovery were that they have a public interest mission to advance knowledge, that research and education are linked, and that the relationship with the federal government is a partnership. Principal investigators in universities proposed research agendas, some of which the government supported through grants. Therefore, it was argued, the university should use other sources of funds to supplement government grants. One historical survey summarized the development of the financial relationship between the federal government and universities:

> At the outset, the federal government provided research funds to universities on terms markedly different from those governing relations with industry. Whereas industrial firms were eligible for reimbursement of full audited costs, universities were permitted to recover only a fraction of their indirect costs. The theory was that since research was a regular function of universities, some of the university's own budget should go to the support of the research performed by its faculty, whatever the source of that support. The earliest NIH reimbursement rate for indirect costs was 8 percent. As federal subvention increased, however, the universities argued that they were in effect subsidizing government in ever larger degrees. In response, the regulations were changed to permit reimbursement of 20 percent of indirect costs and finally, in 1965, by act of Congress, the policy was changed to provide for a negotiated reimbursement of costs, but not full reimbursement.

> The principle adopted was that of "cost-sharing," a notion growing out of the original assumption that some of the charge for university research ought to be borne by the university. (Lakoff, 1978, pp. 173–174.)

When research was a relatively small enterprise, universities could more readily share the costs with the federal government. These modest cost shares did not affect universities' ability to perform their other missions, including teaching and public service. When the research enterprise grew, although universities were willing to continue to share costs, they argued that they could not share costs *at the same rate as when the enterprise was smaller.* The absolute value of the university share in research may have increased, even though the percentage of the share was declining. In a sense, the greater federal share for facilities and administrative costs is a marker of the success of the university-government partnership in research.

There were other changes during this period after World War II. The federal government developed cost principles specific to colleges and universities, recognizing that they perform several distinct but related services. Higher education institutions perform teaching, research, and public service. As noted in Chapter One, each of these activities may share some common resources such as buildings and central management. F&A reimbursement rules provide for how these common costs are allocated to the various functions they support.

GENERAL PRINCIPLES FOR ALLOCATING COSTS

In order to compute an F&A rate for a college or university, costs are divided into three categories, as illustrated in Figure 2.1. We summarize the definitions of the major cost categories here.

- **Direct Costs:** Costs closely tied to a specific project are termed "direct costs." These include salaries for scientists and wages for project team members. In addition to salaries and wages, direct costs also include materials and supplies used in the course of a project. Other direct costs include travel, project-specific equipment, and subcontracts to other organizations.
- **Exclusions:** When projects incur costs for equipment or for payments to subcontractors, these costs must be separated from direct costs. In computing F&A rates, costs for subcontracts over a certain threshold (currently $25,000) and equipment must be separated from direct costs, as shown in Figure 2.1. There are a few other exclusions, but equipment and subcontracts are the most important. The direct costs minus the exclusions are called "modified total direct costs (MTDC)."

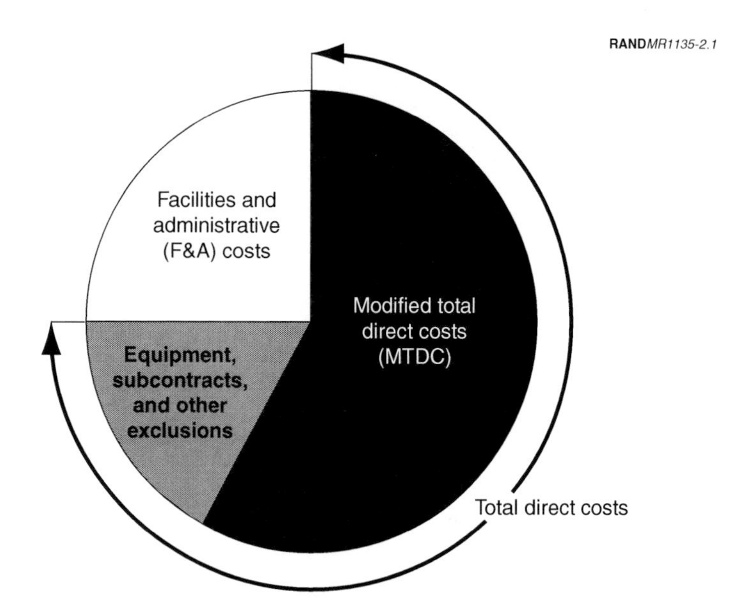

Figure 2.1—Allocating Costs

- **Facilities and Administrative (F&A) Costs:** F&A costs include shared expenses related to facilities or administration of the university. Office of Management and Budget Circular A-21, discussed below, provides definitions for which costs are to be included in facilities and administration.

Facilities costs are

- allowances for depreciation and use of buildings and equipment,
- interest on debt associated with buildings and equipment placed into service after 1982,
- operation and maintenance expenses (such as janitorial, utility, repairs, security, environmental safety, and insurance), and
- library expenses (library operations and materials purchased for the library).

Administrative costs are

- general administration and general expenses (such as central offices for the president, financial management, general counsel, and management information systems),

- departmental administration (including academic deans, faculty administrative work, secretaries, and office supplies),
- sponsored-projects administration (a separate office that administers contracts and grants), and
- student administration and services (operations of student affairs, deans of students, registrar, advising, and counseling), which are normally excluded or limited when computing rates for research.

MTDC, rather than total direct costs, forms the base for calculating F&A costs to projects. The term "base" is also used as a synonym for MTDC because MTDC is the base for distributing F&A costs. MTDC has come to be the accepted base for allocating F&A costs because the direct costs are considered a reasonable indicator of how much benefit the project is deriving from the shared facilities and administration. Because subcontracts and equipment can involve very large expenditures and yet do not necessarily take much advantage of the university's infrastructure, they are excluded from MTDC to compute the base. Other bases are commonly used in nonprofit and for-profit organizations. The discussion below indicates that universities with small volumes of sponsored projects are permitted to use a base of salaries and wages only to allocate shared costs.

As noted, costs for F&A that would be difficult to assign to a specific project are pooled to compute total F&A costs. A university's total F&A rate is computed by dividing F&A costs by MTDC. As the pie chart shows, F&A *costs* may account for only about one-quarter of total costs, but the F&A *rate* is a different number because it is the *ratio* of F&A costs to MTDC. F&A rates of about 50 percent of MTDC are typical of universities. A common misunderstanding is that a 50 percent F&A rate means that 50 percent of total expenditures are for overhead. That is not the case.

If there were no exclusions from direct costs (such as equipment and subcontracts), then all costs would be either MTDC or F&A costs. In that case, a typical 50 percent F&A rate would mean that for each $100 in MTDC, $50 in F&A costs would be allocated. The $50 in F&A costs out of a total budget of $150 means that one-third of project costs are F&A costs in this case. If, in addition, the project incurred costs for large subcontracts or equipment, there would be additional direct costs. In that case, the fraction of project costs for F&A would be lower than one-third. In an extreme case, a grant that funded only equipment purchases would incur zero F&A costs. Based on these cases, a typical 50 percent F&A rate means that between zero and one-third of total project reimbursements are for facilities and administration. In a later section, we use two methods to calculate the average share of F&A reimbursements in total federal outlays for university research.

FORMAL PROCEDURES FOR ALLOCATING COSTS AND NEGOTIATING RATES

Currently, three Office of Management and Budget circulars provide guidelines to federal agencies and research universities for financial management.

- **Circular A-21,** *Cost Principles for Educational Institutions,* establishes principles for determining the costs that apply to research conducted under grants, contracts, and other agreements with universities. This circular distinguishes between direct and indirect costs. Indirect costs fall into two categories: facilities and administration.
- **Circular A-110,** *Uniform Administrative Requirements for Grants and Other Agreements with Institutions of Higher Education, Hospitals, and Other Non-profit Organizations,* provides guidance to grantees and contractors for financial management of federal funds received.
- **Circular A-133,** *Audits of States, Local Governments, and Non-profit Organizations,* creates a vehicle to monitor compliance with cost principles and management regulations.

Although it is the Office of Management and Budget that sets forth cost policies, currently two other agencies are responsible for negotiating F&A cost-recovery rates with universities on behalf of the federal government: the Department of Health and Human Services (DHHS) Division of Cost Allocation and the Department of Defense Office of Naval Research (ONR). According to Circular A-21, the agency (DHHS or DoD) that provides more funds to an institution is responsible for negotiating the institution's rate.

If an institution performs federal projects subject to Circular A-21 totaling less than $10 million (in direct costs) per year, it may use a simplified method (short form) rather than the regular method (long form) to account for its research costs. DHHS and ONR report that (as of June 1999) 685 colleges and universities use the short form. The short form offers two bases for computing F&A recovery rates. The rates may be computed on the basis of just salaries and wages (with or without including fringe benefits in the base) or on a base of modified total direct costs (defined in the discussion of the long-form method below). In addition, short-form rates are computed based on aggregate costs for the entire institution, whereas long-form rates at larger institutions may be computed separately for different functions, such as instruction and organized research. Because of these variations, it is not appropriate to compare rates computed on the short form with those computed on the long form. In this report, we will present information only for long-form schools.

When the cognizant agency and institution begin the F&A rate negotiation, the institution proposes percentages for components of the F&A rate based on data from the institution's accounting system. The process may be formal, with face-to-face meetings, or an informal negotiation conducted through correspondence or teleconferences. The cognizant agency is responsible for formalizing all determinations or agreements with an institution. All of ONR's negotiators are located at its headquarters. DHHS, on the other hand, locates negotiators in four regional offices.

DHHS and ONR report that, as of June 1999, 282 institutions use the long form. Rates for larger institutions using the long form are generally determined by major function: e.g., instruction, organized research, public service, and patient care. Thus F&A costs applicable to instruction are pooled separately from those applicable to organized research, resulting in different F&A rates for these two major functions. For long-form institutions, F&A costs must be apportioned on the basis of modified total direct costs (MTDC). MTDC includes most direct costs of projects: salaries, wages, fringe benefits, materials and supplies, and travel. MTDC also includes the first $25,000 of each subcontract. Subcontract amounts over $25,000 per subcontract are excluded from MTDC. Equipment, capital expenditures, and certain other expenses are also excluded from MTDC.

The cognizant agencies are permitted to negotiate several types of single-year or multiyear rates (defined in Appendix B). Agencies now prefer to use predetermined rates based on information from a base period. These predetermined rates are typically in effect for two to four years and are not subject to changes during the agreed period. Predetermined rates reduce costs of negotiating rate agreements and allow all parties to budget more precisely during the predetermined period. When a predetermined rate is not used, a negotiated fixed rate with a carry-forward provision may be used. Under this type of rate, any differences between the estimated costs used to establish the fixed rate and actual costs during the period are carried forward to a subsequent period as an adjustment.

Universities perform at least three major functions: instruction, organized research, and public service. Many universities operate medical schools, which perform other functions, such as patient care. It can be complex to allocate the costs of using and operating facilities that several of these functions may share. Similarly, administrative services are shared across multiple functions and must also be allocated. In a long-form negotiation, the university uses allocation methods prescribed in Circular A-21 to divide total facilities and administrative costs among the various functions and organizations within the university. It is up to the university to propose what the annual costs are for components of both facilities and administration. Universities may rely on special studies to determine an allocation in the case of some cost elements. For instance, research laboratories often incur higher costs for utilities than instructional space does. Before 1999, universities that wished to claim these higher costs had the option to perform special engineering studies to allocate the utility costs to research space. Starting in 1999, these special studies were eliminated. Universities whose cognizant agencies had approved special studies instead were allowed to claim a flat amount of 1.3 percent of MTDC (1.3 points on their F&A reimbursement rate) in lieu of the special studies. This 1.3 percent of MTDC represented the average costs documented through the former special studies for utility use in research space.

In the interest of concluding their negotiation and reaching agreement, both the university and the cognizant agency may decide not to pursue claims they feel might be justified under Circular A-21. If a university chooses to exclude a certain cost element from its proposal, it waives the right to recover costs for that element. In addition to such compromises made during the negotiation process, universities may deliberately omit a cost element because they do not wish to incur the expense to document those costs or because they prefer to maintain an F&A rate that is competitive with peer institutions. Some observers believe that government negotiators seek, in some cases, to maintain comparability among institutions. An institution with a different cost structure may therefore be discouraged from including certain costs in its negotiated rate. For these reasons, the negotiated F&A rate may represent less than the full share of F&A costs attributable to federally sponsored research.

So universities might not recover the full costs attributable to federal projects because negotiated rates may be set below actual costs. Further, universities do not necessarily recover the full amount of the negotiated F&A rates. Statutory requirements and agency policies limit recovery of F&A costs on certain grants to fixed levels. For example, these limits are imposed on all Department of Agriculture research grants, specific NIH grants for predoctoral and postdoctoral training, and certain Department of Education grants. Even where no limits apply to F&A costs particularly, many agencies require or expect universities to share some of the costs of a research project.

CALCULATING THE F&A PORTION OF FEDERAL RESEARCH OUTLAYS

How is the share of federal research dollars devoted to facilities and administration calculated? We can approach the calculation in two ways. One method is to examine the projects conducted by universities with federal funds to compute the quantities in Figure 2.1 above. The other method is to examine federal agency outlays to compute the fraction of awards that pay for F&A costs. Both methods are subject to incomplete data and hence can only produce estimates.

For the first method, we use data collected by the Council on Governmental Relations (COGR), an organization for research universities that deals primarily with federal administration of sponsored programs. COGR has conducted an annual survey related to F&A costs in its member institutions, which include most major research-intensive universities. We can use these data in the first method of computing the F&A portion of federal research outlays. In the most recent data, which cover the 1998 fiscal year, 128 higher education institutions reported data. Some of the institutions did not report complete data on the items we need for this analysis, so we were able to use 102 complete records.

For projects with federal sponsorship, these 102 institutions reported a total of $4.5 billion in MTDC and $1.2 billion in exclusions from MTDC (equipment, subcontracts, and other). They reported receiving $1.9 billion in F&A reimbursement from the federal government for these projects. The MTDC and F&A figures are totals from the survey reports, whereas the figure for exclusions from MTDC involves an estimate.[1]

These figures indicate a breakdown of 75 percent for total direct costs (59 percent for MTDC and 16 percent for exclusions) and 25 percent for F&A costs. That breakdown is plotted in Figure 2.2 using the schema from Figure 2.1.

Using the negotiated rates for institutions in this survey, we compute that the (weighted) average negotiated F&A rate for these institutions is 51.2 percent of MTDC. This 51.2 percent does not mean that more than half of all payments go for F&A costs. The previous paragraph explains that 25 percent of total outlays are for F&A costs.

If these institutions received their full negotiated rates for each federally sponsored project, they would have received a total of $2.3 billion for F&A costs, rather than the $1.9 billion they actually did receive. Therefore, $0.4 billion of the negotiated costs for F&A at these institutions did not get reimbursed by the federal government.

We would like to compute these quantities for all higher education recipients. The total of the federal payments reported on the survey (using our estimate for

---

[1]The survey requests separate figures for the federal share of MTDC and F&A costs but does not ask for exclusions from direct costs to be broken down between federal and other. We make the assumption here that exclusions from direct costs can be allocated in proportion to the MTDC for federal and nonfederal sponsors. There may be systematic reasons why federal projects include more or less of these exclusions than other projects. As a result, this breakdown could be off by a few percentage points.

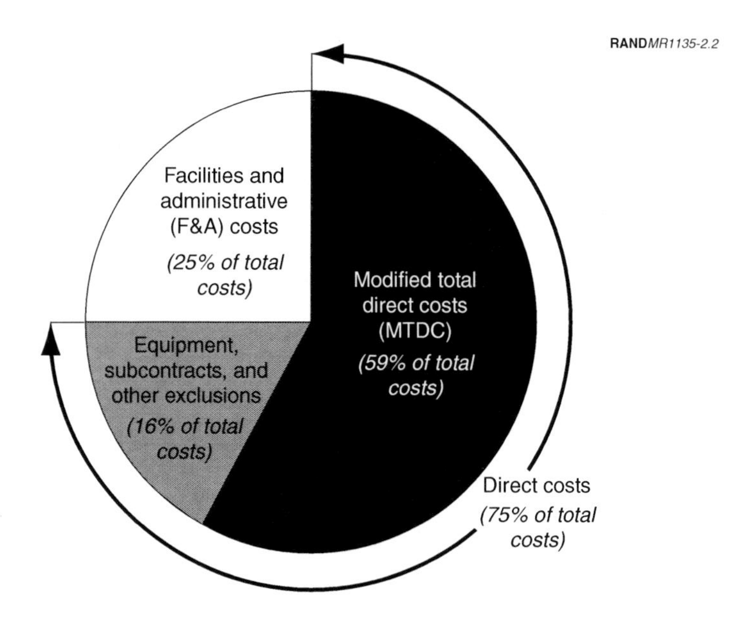

SOURCE: 1998 COGR Survey. Complete data on 102 institutions.

Figure 2.2—Average Distribution of Federal Awards

exclusions) is $7.6 billion, or a little more than half of all federal research payments to higher education that year. We are missing data on the experience of the universities and colleges that make up the other half of federal outlays. It is difficult to say whether the percentages that we are able to calculate from these data would apply to the institutions not represented here. In particular, the experience of other institutions in recovering F&A costs from the federal government may be different. The available data suggest that institutions with smaller federal research programs recover less of their negotiated F&A costs than institutions with larger federal programs. Because the COGR data include more of the large recipients and fewer of the small recipients, the institutions omitted from this survey may tend to recover less of their negotiated F&A rates, on average. There are other sources of uncertainty, however. Because of the estimation for exclusions, there may be a slight error in the breakdown in Figure 2.2. In any case, we do not know if the breakdown in Figure 2.2 is applicable to the omitted institutions.

In the entire system of universities and colleges, federal agencies awarded $15.1 billion for research and development at higher education institutions in fiscal year (FY) 1997.[2] Projecting the breakdown in Figure 2.2 onto the omitted insti-

[2]Federal Obligations for Sciences and Engineering, FY 1997. NSF WebCASPAR system.

tutions, we would calculate that of the $15.1 billion in federal outlays, $11.3 billion (75 percent) is for total direct costs and $3.8 billion is for F&A costs. Continuing this projection, universities and colleges as a whole were not reimbursed for $0.8 billion of negotiated F&A costs for federally sponsored research.

The second method of examining federal outlays is with agency data. NIH maintains statistics, discussed in Chapter Four, for research awards to universities that are not subject to any agency-specific limits on F&A reimbursements.[3] For those awards, in 1998, 31 percent of the total funds were for F&A costs.

As noted above, some agencies have statutory caps that limit the amounts they pay for F&A costs. NIH accounting systems record that in FY99, the latest year available, universities received $592 million in MTDC for programs with statutory or regulatory limitations on F&A recovery. These programs include graduate and postdoctoral training grants and career awards, but not individual fellowships. Universities received $46 million in F&A costs on those grants, which limit F&A costs to 8 percent of MTDC. If the average rate of 51 percent of MTDC applied to these grants, F&A costs would have been $302 million. Therefore, about $256 million in otherwise allowable F&A costs were not reimbursed because of the limitations on NIH programs. Including the effect of these programs, we estimate that of all NIH awards to higher education, F&A costs represent about 29 percent of total awards.[4]

The Department of Education observes the same limitations on its training grants. Because, in the field of education, most of those grants are considered part of dissemination rather than research training, they are not included in the research budget. Despite the difference in classification, universities must still make up the unreimbursed amounts from their resources, since, according to Circular A-21, these costs cannot be recovered from other projects. A few Department of Education programs subject to F&A limitations do fall within the research budget, but we did not calculate the unreimbursed amount.

The Department of Agriculture (USDA) has a number of limits on F&A reimbursements for its competitive research grants. Most programs limit F&A reimbursement to 14 or 19 percent of the total award costs. As a result of these lim-

---

[3]Governmentwide limits on F&A reimbursement still apply. As discussed in Chapter Five, a 26 percent cap applies to administration for all higher education institutions. In addition, we are aware that NIH must impose a salary cap of $136,700 and cannot reimburse salaries, especially of physicians, above that level. F&A costs for salaries above the cap are also not reimbursed.

[4]NIH data for FY 1998. In FY 1998, NIH made $9.304 billion in awards for research projects and centers, which are generally allowed full F&A reimbursement. NIH made $0.659 billion in training and faculty career awards, which are allowed 8 percent of MTDC for F&A. Considering exclusions from MTDC, F&A represents 6.6 percent of these total awards. The weighted average, using 31.0 percent for research projects and 6.6 percent for training and career awards, is 29.4 percent.

its, in 1998, 13 percent of USDA research project funds were expended for F&A reimbursements.[5] USDA, according to its internal tracking, reimburses about half of the negotiated F&A costs on its research grants. Universities are contributing about $11 million per year from their funds.

The bulk of USDA's awards classified as research and development is not for competitive research project grants. These awards are largely congressional formula funding for agricultural extension services, which involve extensive cost-sharing with the states. At least half of the costs of agricultural extension programs are generally borne by the states or other nonfederal sources. In the face of such extensive cost-sharing, there is no clear way to assign which funds are paying for the F&A costs of agriculture extension. The federal side considers all F&A costs paid from the state share. If we accept this view, then USDA's overall research and development budget contains nearly zero reimbursement for F&A costs, because the only F&A costs attributed to the USDA budget are for the relatively small research project grants described above.

Other agencies also have required cost sharing. For NSF, recipients must share one percent on all grants. Some universities may treat this required cost sharing as coming from the F&A portion of the budget; others may account for it as direct costs; others may share it in proportion to the total budget in both categories. Voluntary cost sharing may likewise appear under F&A or direct costs.

Because NIH awards the majority (about 51 percent) of all federal research funding to higher education, its award structure heavily influences the average. Except for USDA, other major funding agencies are likely to resemble NIH in their experience, although the awards they make and the universities they support may differ. Because the USDA's share for F&A costs is so low, the average for all agencies is probably close to, but smaller than, the NIH percentage. Based on the evidence we examined, a reasonable estimate of the true fraction of federal outlays for F&A costs is in the range of 24 to 28 percent.

The figure of 25 percent based on the data from universities accords very well with this range. As noted in that discussion, the 25 percent figure may not be quite accurate if applied to all institutions. It may be in error by a few percentage points because the institutions reporting in the COGR survey differ in some ways from those not participating in the survey. If we accept the 25 percent figure as applicable to the whole set of higher education institutions, we conclude that, of the $15.1 billion in federal outlays, 25 percent, or about $3.8 billion of

---

[5]Data provided by USDA for FY 1998. That year, the USDA made awards of $84,176,009 for competitive research grants. Of this amount, $10,910,782 was for F&A costs. The ratio of F&A costs to total awards was 13.0 percent.

this amount, was for F&A reimbursements. The direct costs of projects were 75 percent of total outlays, or about $11.3 billion.

As noted above, actual reimbursements for F&A costs on some projects are subject to limits. The full negotiated amounts for F&A are not reimbursed for those projects. We now perform a series of calculations to estimate unreimbursed F&A costs. These calculations are necessary because we are estimating the F&A costs associated with the current set of research projects pursued in higher education with federal funding. Since we do not know the exact split of direct and F&A costs in the current outlays, we first estimate that split using available information. The direct costs identified in this way represent the current set of projects. Using estimates of the split of direct and F&A costs that occurs when no special limits are imposed, we can identify the approximate F&A costs that match the calculated direct costs. By comparing this calculated amount to the amount of F&A costs in current outlays, we compute unreimbursed F&A costs.

The NIH data provide one estimate of project budgets when most special limits do not apply. For NIH research project awards, 31 percent of total outlays reimburse F&A costs. If full F&A reimbursement had applied to every grant and contract with universities and colleges from every agency and program been accorded, direct costs would be 69 percent of total costs and F&A costs would be 31 percent. The actual figure might be a few percentage points different, to the extent other agencies work with different universities than NIH and experience different cost structures.

Using a split of 75 percent direct costs and 25 percent F&A costs, we estimated above that direct costs were $11.3 billion. Under these assumptions, with direct costs of $11.3 billion, negotiated F&A costs would have to total $5.1 billion in order to make total expenditures $16.4 billion, split in the proportion 31 percent for negotiated F&A expenditures and 69 percent for direct expenditures. We estimated above that the federal outlays for F&A costs were $3.8 billion, meaning that about $1.3 billion was not provided by federal funds. Under these assumptions, the federal government appears to reimburse about 75 percent of F&A costs attributable to federal projects, based on negotiated F&A rates.

We do not simply split the actual outlays of $15.1 billion in the proportion 31 to 69 percent. That procedure would represent a change in the projects pursued at universities, since we would be eliminating direct costs that are currently expended on federally funded research projects. Our calculations here hold the current set of projects fixed in order to calculate how costs are shared between federal agencies and universities.

Because these figures are based on assumptions as well as actual data, they are not precise. We can get a measure of the uncertainty by varying the parameters. If we use a low-end figure of 24 percent of federal outlays to represent F&A costs, we would find that reimbursed F&A costs were $3.6 billion and unreimbursed costs were $1.5 billion. In this case, the federal government appears to reimburse about 70 percent of the F&A costs. If we use a high-end figure of 28 percent of federal outlays to represent F&A costs, we would find that reimbursed F&A costs were $4.2 billion and unreimbursed costs were $0.7 billion. In this case, the federal government appears to reimburse about 87 percent of the F&A costs. Given the uncertainty involved in these figures, it is appropriate to round this 87 percent figure to an even 90 percent, making the range we estimated 70 to 90 percent of negotiated F&A costs reimbursed. All of these figures are based on the assumption that the 31 percent of project costs for NIH represents the full federal share when negotiated rates are used without limitations.[6]

Extrapolating from the COGR university data above, we estimated that universities as a whole are not reimbursed for about $0.8 billion in F&A costs. This figure is at the lower end of the range we estimated based on the government data.

To summarize, using the data we have available and making assumptions where data are inadequate, we estimated that federal outlays for research in higher education include about 25 percent (24 to 28 percent) for F&A costs. Federal F&A reimbursement does not cover the full negotiated federal share of university F&A costs. Using university-reported data, we estimated that roughly $0.8 billion of F&A costs for federal projects was not reimbursed. Using the data from federal agencies, we estimated that this amount was between $0.7 and $1.5 billion. Because the data do not cover some important segments of agency funding or institutions, we cannot be more precise than the ranges we present here.

We can identify some reasons for F&A costs not being reimbursed. The NIH training and career awards discussed above require universities to fund about $250 million per year in F&A costs that these grants do not reimburse. Awards from the Department of Education and USDA also require university funding for some F&A costs. The remainder of the unreimbursed F&A amounts are primarily general cost sharing, but mandatory and voluntary, which universities may consider either direct or F&A costs.

To compensate for underreimbursement of negotiated rates, universities might pursue several strategies. Although we might be concerned that universities would shift costs from projects with less generous agencies to projects with

[6]In another analysis using earlier data from the COGR survey, analyzed in Chapter Four, we estimate that institutions recover about 77 percent of their negotiated F&A rates. This is consistent with the range we estimate here.

more generous agencies, OMB Circular A-21 indicates that its controls prevent this form of cost shifting. Circular A-21 requires that facilities and administrative costs be apportioned to all organized research projects regardless of whether full—or any—reimbursement is available from federal sources. The methods of apportionment in Circular A-21, such as allocating by square footage used in each activity, are not amenable to shifting from the true use to another reported use. So it can be assumed that universities do not have latitude to assign costs to more generous agencies in preference to less generous agencies.

Beyond these statements, we do not have data to indicate how universities compensate for underreimbursement. Many possibilities exist for how universities fund costs that are not reimbursed. They may use private gifts or endowment income, state appropriations, or other sources of revenue. We cannot be precise about the mechanisms used because of the overlap of university missions and funding sources.

COST-SHARING

The calculations above indicate that universities share facilities and administrative costs in the range of several hundred million dollars per year or more. In addition to agency-specific limits on facilities and administrative reimbursement, there are other cost-sharing requirements. The NSF, for example, has a statutory requirement that universities provide some cost-sharing on all projects, at a minimum of one percent of the total project cost (including direct and F&A costs). Individual programs within agencies may seek additional cost-sharing, or universities may voluntarily propose higher cost-sharing.

In 1991, OMB Circular A-21 was revised, in response to congressional interest, to place a cap on the level of administrative costs that could be included in rate negotiations. From this point, universities could include only 26 percent of MTDC for administration. Any administrative costs over this amount would not be included in the rate negotiations and hence not reimbursed. If universities continued to experience costs in excess of this amount, they would have to pay for them from other funds.

As mentioned in Chapter One, universities pursue several functions simultaneously and with shared resources. Universities must account for faculty effort and the use of shared facilities in order to allocate these shared costs to the appropriate functions. One function is instruction—the teaching of students. Another function is organized research, which includes sponsored projects as well as any separately budgeted research activity, even if paid for with the university's own funds.

Not all research activity is part of an organized research project, though. Small projects and general scholarly work without external funding are considered part of basic faculty workload. This work, termed "departmental research," is considered part of the instructional function. Effort on both instruction and departmental research is combined for purposes of allocating shared costs.

If faculty time is contributed to a general line of research but not shared on the budget of a funded project, then it may be considered departmental research. The university pays for the direct costs of departmental research as well as any associated facilities and administrative costs, just as it does for instruction (which is combined with departmental research for accounting).

If faculty time is formally shared on a project budget, that time must be accounted for as part of organized research, even though it is not sponsored by the government. The university must bear the costs not paid by the government—both direct costs and associated facilities and administrative costs. A university's cost share on a federal project is supposed to be counted as part of the MTDC base in calculating F&A rates, even though it is not reimbursed. Universities that contribute more in cost-sharing for research will see lower F&A rates because they have a larger base. When F&A costs are spread over a larger base, the F&A rate is reduced, resulting in lower F&A rates for universities with more cost-sharing.

The question may be raised about whether universities can shift costs onto organized research rather than departmental research, which is a component of instruction. Because F&A costs are allocated to both instruction (including departmental research) and organized research, universities do not appear to have the ability to shift them from departmental research to federal projects. Federal projects benefit from faculty effort provided under departmental research in a similar way to faculty effort provided as a cost share under organized research.

As mentioned in the introduction, universities and their faculty are voluntary participants in this system. They offer and provide these cost shares because they perceive good reasons for them. Specifically, federal projects bring prestige to faculty in their careers and universities as institutions.[7] As Figure 1.1 shows, universities provide much of research funding from their own sources. Some of these funds support entire projects that do not have outside sponsorship. Other funds support part of projects funded with federal grants or contracts. The funds support direct costs and F&A costs.

---

[7]Several researchers have examined the link between sponsored research and prestige, including Fairweather (1988), McGuire et al. (1988), and Grunig (1997).

EXAMINING F&A RATES

With this background, we now turn to the six questions raised by Congress. As stated in the introduction, this report analyzes available data related to each of the questions. We first compare facilities and administrative rates in higher education with those in other sectors. Then we examine F&A rates within higher education, the impact of changes in Circular A-21, and the impact of federal and state law. We next examine options to reduce F&A reimbursement rates and options for creating a database. A conclusion brings together some of the insights from the analyses.

Chapter Three

ISSUE 1: COMPARISON OF F&A RATES ACROSS SECTORS

In the course of the policy debates on this subject, some have expressed concern about how the cost structure of universities compares with other enterprises that do business with the federal government. This chapter summarizes available information on that question.

Universities, commercial enterprises, government laboratories, hospitals, and other organizations differ in important ways. Each of these types of enterprises is governed by a different set of federal regulations for grants and contracts. Because of these different federal regulations and other business factors, different accounting conventions are used in these sectors. In addition, institutions in these sectors have different functions, scope, and scale. These differences influence how F&A rates are computed, making for potential misunderstandings when comparing rates from one sector with another. For example, the base used to distribute F&A costs in universities is modified total direct costs (MTDC), whereas different bases are used in the other sectors. As a result, a given rate number has different meanings in the different sectors. In this chapter, we look at data that attempt to place several sectors on the most similar base possible for purposes of comparison.

In Chapter Two, we analyzed federal reimbursements for F&A costs. In this chapter, we consider true total project costs rather than reimbursements. True costs are computed based on the total costs of operations as reported in organizational accounting systems, regardless of whether the federal government would reimburse those costs. For example, all administration is included, even though universities can recover only 26 percent of MTDC for administration on federal agreements. The available evidence indicates that the fraction of true costs in universities that are F&A costs is generally comparable with or somewhat smaller than indirect costs for other performers of research. In all the sectors studied, F&A costs accounted for about one-third of true total costs. For the universities, F&A costs are about 31 percent of total project costs. In Chapter Two, our analysis concluded that average federal agency payments include between 24 and 28 percent of funding for F&A costs. Cost-sharing by universities accounts for the difference between these lower figures (24 to 28 percent) for payments and the 31 percent figure for total costs.

COMPARING RESEARCH COSTS IN UNIVERSITIES, FEDERAL LABORATORIES, AND INDUSTRIAL LABORATORIES

Although data are maintained on F&A rates for universities, no comparable government or private data exist for commercial enterprises. Corporate indirect cost rates are considered proprietary information that companies are sensitive about disclosing, because these rates have important effects on competitiveness. To our knowledge, government agencies do not maintain a centralized registry of corporate indirect cost rates.

One recent attempt was made to compare F&A costs on a similar basis across sectors. In 1996, on behalf of the Government-University-Industry Research Roundtable, Arthur Andersen, LLP, conducted a study examining whether direct and indirect costs varied across research organizations in three sectors: universities, federal laboratories, and private companies. Arthur Andersen solicited the participation of seven universities, 13 federal laboratories, and 13 industrial (for-profit) laboratories. The selection of participants was based on availability and willingness to cooperate. Because only a few institutions for each type were examined and they were not selected to represent all institutions of that type, we cannot say how these results would differ for a larger, more comprehensive group of institutions. The data used by organizations in the study were from an available recent fiscal year between 1991 and 1994.

The study assessed total costs of performing and supporting research, without regard to the amounts actually recoverable from the federal government for research. The study concluded that the division of costs is similar for all three sectors. In each sector, Table 3.1 shows that about one-third of costs were classified as indirect and about two-thirds as direct. As a fraction of total costs, universities had the lowest percentage classified as indirect (31 percent). Federal laboratories were somewhat higher at 33 percent and industrial laboratories were higher still at 36 percent.

One reason nonprofit research institutes and industrial laboratories may have slightly higher indirect costs is that these institutions must allocate certain central organization costs (such as the cost of facilities and senior management) purely to research, because this is their only function. Universities, on the other hand, can distribute these central organization costs to instruction, research, and other functions. Research, then, may bear a somewhat smaller central organization cost in a university.

Table 3.1

Fraction of Total Costs Classified as Direct and Indirect in Three Sectors (Arthur Andersen, 1996)

	Universities (n = 7)	Federal Labs (n = 13)	Industrial Labs (n = 13)
Costs classified as direct	69%	67%	64%
Costs classified as indirect	31%	33%	36%

SOURCE: Arthur Andersen, 1996.

Considerable university data preceded the imposition of the administrative cap (discussed in Chapter Five) that took effect after 1993. Universities likely made some changes in their administrative operations after the cap was imposed, reducing spending on administration. As a result, university indirect costs may be even lower now than in the older data used for this study. In any case, as described in Chapter Two, the administrative cap means that fewer administrative costs are *reimbursed* today. Because nonprofit and commercial research institutes are not subject to this sort of cap, universities may well have fewer costs for administration. Even if their costs for administration were the same as for the other sectors, they would receive less federal reimbursement for them.

COMPARING BIOMEDICAL RESEARCH COSTS IN UNIVERSITIES, HOSPITALS, AND RESEARCH INSTITUTES

As noted in Chapter One, NIH accounts for more than half of all federally sponsored research in higher education. NIH has compiled statistics on F&A reimbursements in response to congressional requests. These reports provide us with another way to compare cost structures across sectors.

The Department of Health and Human Services (1991) found that F&A rates vary by region, possibly stemming from differences in climate (and hence utility costs) or labor costs. To compare projects across sectors more accurately, NIH compiles reports for Congress using geographical region to group statistics on F&A rates. The latest report was issued in February 1999 and covered NIH-funded research projects by performers in four sectors. It showed that F&A costs were proportionately highest in the Northeast and lowest in the South and Midwest. Table 3.2 summarizes the information from the 1999 report. Overall, universities were awarded 31 percent of their total awards for F&A costs on NIH projects. (This is distinct from the governmentwide average of 24 to 28 percent computed in Chapter Two.)

This study included only grant programs for which full negotiated F&A rates are allowed (with the administrative cap and salary cap in effect). Training grants

Table 3.2

Indirect Cost as a Percentage of Total Cost Awarded (NIH, 1999)

Region	Higher Education	Hospitals	Research Institutes	For-Profit Small Business	All Sectors
Midwest	30.0	26.4	26.7	19.1	29.5
Northeast	33.3	31.2	34.5	20.3	32.9
South	29.4	23.6	36.6	28.8	29.3
West	29.0	29.6	36.8	25.1	30.8

SOURCE: Office of Extramural Research, NIH, Department of Health and Human Services, February, 1999

received by many universities strictly limit F&A reimbursement and are excluded from these calculations. As a result of these limitations, F&A costs calculated over all NIH awards would show a lower percentage than the figures in Table 3.2. The table indicates that for the NIH awards covered, higher education institutions received between 29 and 33 percent of awards for F&A costs, depending on the region of the country. Awards to hospitals in two regions (Northeast and West) were similar in composition to those for higher education. In the other two regions, hospital awards were somewhat lower for F&A costs. Research institutes received higher fractions of awards as indirect in three of the four regions compared with higher education.

Although Table 3.2 indicates that F&A percentages for for-profit small businesses were smaller than for higher education, these awards represent only a limited portion of the relationship between federal agencies and for-profit organizations. The awards summarized in this analysis are those made under two small business set-aside programs, the Small Business Innovation Research (SBIR) Program and the Small Business Technology Transfer (STTR) Program. Projects under the SBIR and STTR are funded for direct costs, indirect costs, and a fee of up to 7 percent of total costs. Fees are not included in the amounts used to calculate the indirect cost percentages in the table.

Although the guidelines for these grants call for indirect cost reimbursement, some grantees may not propose full indirect costs, or any at all. These programs provide assistance to small businesses; therefore, the businesses themselves may share in the costs of research as part of their overall research and development efforts. Guidelines for Phase 1 grants under both SBIR and STTR suggest a maximum budget of $100,000 (although this is not a firm limit at NIH). A number of grantees propose projects that would exceed $100,000 and can justify funding of $100,000 with direct costs alone, so they do not include indirect costs in their proposals. Therefore, the actual level of indirect costs for these projects is higher than reported in Table 3.2. Guidelines for Phase 2 grants suggest a $750,000 maximum budget, and budgets generally include indirect costs.

In contrast to these programs, most federal payments to for-profit organizations are in the form of contracts. An examination of indirect costs on contract payments to for-profit firms was not possible, but observers agree that their indirect costs on federal contracts for similar research functions are generally higher than in universities. The Arthur Andersen study of the costs of research corroborates this view because for-profit laboratories had a 36 percent share of total costs as indirect, compared with 31 percent in the universities studied.

Overall, these comparisons are only broadly indicative of cost structures, because they are premised on comparing quite different organizations with different accounting regulations and reimbursement structures. What evidence is available indicates that the fraction of awards to universities that pays for F&A costs is generally comparable with or somewhat smaller than indirect costs for other performers of research. The true share of costs for facilities and administration appears to be about 31 percent of total costs. The analysis in Chapter Two indicates that federal payments to universities are about 24 to 28 percent for facilities and administration; hence the federal government is not reimbursing the true costs attributed to federal projects.

Chapter Four

ISSUE 2: DISTRIBUTION OF F&A RATES BY SPENDING CATEGORY

The most recent data show that the negotiated rates for facilities account for about half of total negotiated F&A rates. The administrative component accounts for the other half. Since 1988, total negotiated F&A rates have been basically level, with some shift away from the administrative component and toward the facilities component. Within the facilities component, there are different patterns for its three subcomponents: infrastructure, operations and maintenance, and library. Among these subcomponents, negotiated rates for infrastructure have increased, while the rates for the other subcomponents have decreased. Although the cap on administration narrowed differences in F&A rates between public and private universities, the differences remain noticeable. Private universities, on average, have F&A rates about 10 percentage points higher than public universities. The difference in rates between private and public universities appears to arise from several factors. Private universities may have more-expensive facilities. In any case, private universities use depreciation accounting whereas many public universities still rely on use allowances, which may result in lower facilities rates. On average, private universities have greater incentives to recover F&A costs than do their state-supported counterparts, because they do not have state appropriations.

DATA AND MAJOR TRENDS

To analyze recent trends in F&A total rates and components, we requested data from ONR and DHHS on the institutions they regularly track. ONR tracks all of its long-form universities—currently numbering 24. (For a review of the long form, see Chapter Two.) DHHS has tracked 118 of the largest research universities, all using the long form (plus a small number that have recently transferred from ONR). In total, DHHS and ONR data cover between 145 and 153 universities. When we analyzed the data, we did not include institutions with only a provisional rate during a given year. As a result, we have a slightly different number of institutions with regular rate agreements in each year.

There are two basic ways to present F&A costs and rates. In Chapters Two and Three we used one basis: the fraction of total awards or project costs for F&A costs. We shift from here on to presenting negotiated F&A rates as a percentage of MTDC. As explained in Chapter Two, negotiated rates may vary from the actual cost structure of the university because of mandatory and voluntary cost-sharing. At the end of this chapter, we introduce additional data on the actual recovery rates of universities.

In discussing negotiated rates as a percentage of MTDC, it is useful to recall the relationship between negotiated rates and the fraction of total federal payments that goes for F&A. The average negotiated F&A rate for the universities analyzed in Chapter Two is 51 percent of MTDC. That 51 percent figure does not mean that half the expenses are for facilities and administration. The analysis in Chapter Two shows that federal outlays to those universities were divided in the proportion 25 percent for F&A reimbursements and 75 percent for direct costs.

Table 4.1 and Figure 4.1 summarize the average negotiated facilities and administrative reimbursement rates as a percentage of MTDC for this group of institutions from 1988 through 1999.

The average F&A rate has changed very little between 1988 and 1999. There has been some shift in the major components of the rate, though. Administrative costs reimbursed through F&A rates have declined, while facilities costs reimbursed have increased. We do not know if true administrative costs at universi-

Table 4.1

Average of Total Negotiated F&A Rate and Major Components (percentage of MTDC)

Fiscal Year	Number of Institu-tions	Total Rate	Adminis-trative	Facilities	Carry Forward	Lowest Total Rate	Highest Total Rate
88	146	50.6	27.3	23.2	0.1	25.0	87.5
89	146	50.8	27.6	23.6	–0.3	25.0	82.3
90	146	51.8	27.7	24.1	0.0	33.5	78.0
91	146	51.9	27.5	24.3	0.1	37.0	88.0
92	149	52.2	27.4	24.8	0.1	37.0	88.0
93	147	50.5	24.9	25.6	0.1	33.5	83.0
94	147	50.6	25.1	25.7	–0.1	33.0	83.0
95	148	50.9	25.1	25.8	0.1	36.3	79.9
96	148	50.8	25.1	25.6	0.1	36.1	79.9
97	148	50.9	25.3	25.9	0.1	25.0	79.9
98	153	51.0	25.2	26.0	–0.1	37.1	79.9
99	145	50.8	25.2	25.4	0.0	34.9	74.5

SOURCE: Database compiled from ONR and DHHS, 1999.

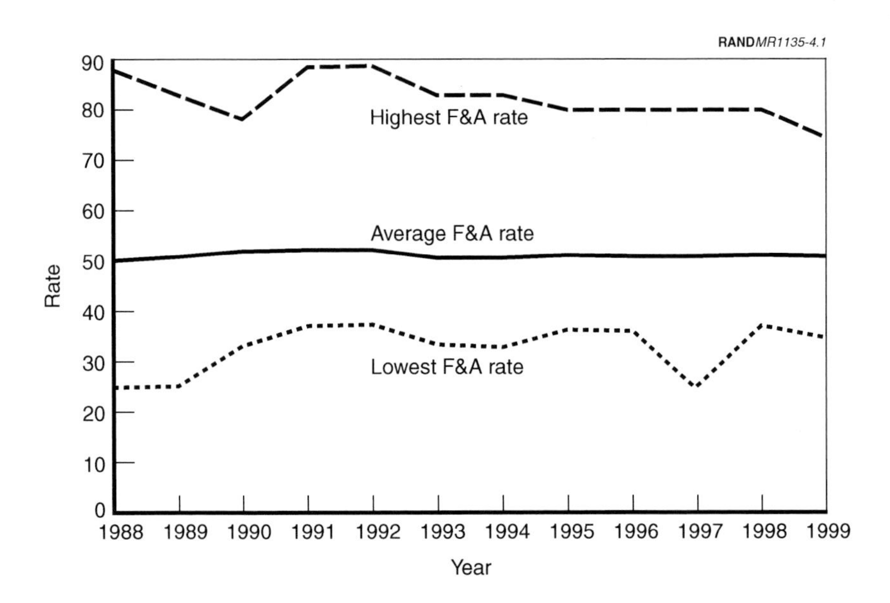

SOURCE: Database compiled from ONR and DHHS, 1999.

Figure 4.1—Trends in Average, Highest, and Lowest Negotiated F&A Rates, 1988–1999 (percentage of MTDC)

ties have decreased, because in 1993 universities became subject to a 26 percent cap on reimbursement of indirect costs. The administrative cap is discussed further in Chapter Five. Because of the 26 percent administrative cap, it is not useful to distinguish subcomponents of the administrative portion of the rates, such as general administration, sponsored projects administration, departmental administration, and student services. Because many universities present true administrative costs that exceed the 26 percent cap, it is not clear exactly which costs the federal government is actually reimbursing. If universities did not reduce their administrative costs following imposition of the cap, they are using other sources of funds to pay for the costs that exceed the cap.

In contrast to the administrative components, facilities components offer some information about trends. Table 4.2 summarizes the average values for these components. The column labeled "Infrastructure" includes negotiated rates for buildings and equipment at a university. The total negotiated facility rate increased modestly from 1988 to 1999. The larger total arises from increases in the infrastructure portion, while operations and maintenance budgets have decreased. The trends in these facilities components are discussed further in Chapter Five.

Figures 4.2 and 4.3 graph the overall trends in negotiated F&A rates during the same period. The strong effect of the administrative cap in 1993 is evident in Figure 4.2. Since then, the administrative component has been flat.

Table 4.2

Average of Facilities Components of Negotiated F&A Rates (percentage of MTDC)

Fiscal Year	Facilities Total	Infra-structure	Operations and Mainte-nance	Library
88	23.2	5.9	15.2	2.1
89	23.6	6.1	15.2	2.2
90	24.1	6.5	15.4	2.2
91	24.3	7.0	15.1	2.2
92	24.8	7.3	15.5	2.1
93	25.6	7.9	15.6	2.0
94	25.7	8.2	15.5	1.9
95	25.8	8.3	15.6	1.9
96	25.6	8.4	15.4	1.9
97	25.9	8.6	14.8	1.8
98	26.0	8.8	15.2	1.9
99	25.4	9.0	14.7	1.8

NOTE: Infrastructure includes depreciation, use allowances, interest, and other costs.

SOURCE: Database compiled from ONR and DHHS, 1999.

Figure 4.3 shows that negotiated rates for facilities increased from 1988 to 1993 and have been basically unchanged since 1993 with some slight fluctuations.

VARIATIONS IN NEGOTIATED F&A RATES

Within these averages, there is considerable variation from one institution to another. Several studies have analyzed the reasons for these variations. Different regions of the country have different climates with corresponding requirements for heating and cooling. Universities with more modern facilities will generally have higher expenses for infrastructure (including depreciation and interest), although expenses for operations and maintenance may be lower. The size of the institution's research base may have an impact on F&A rates because more activity within a given space enables the university to operate more efficiently. Even more important is a good match between the size of the research base and facilities to conduct the research at maximum efficiency. Facilities that are underutilized—or too crowded—are not good values for the university or the government. A good match makes it more likely that the government is reimbursing to the university at a level that closely matches its costs to operate facilities. [1]

---

[1]These and other factors are discussed in NSF, 1991; COGR, 1988; and COGR, 1998b.

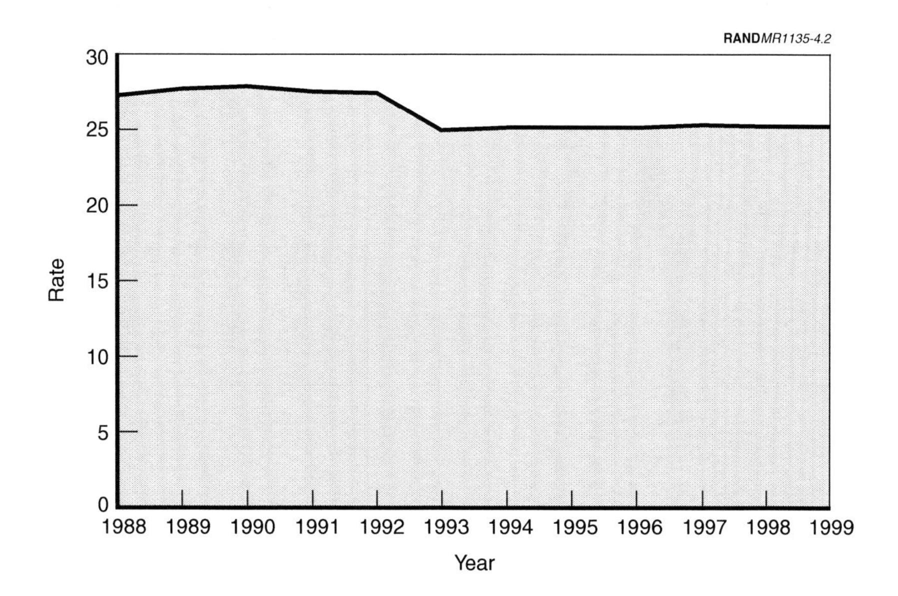

SOURCE: Database compiled from ONR and DHHS, 1999.

Figure 4.2—Trends in Administrative Component of Negotiated F&A Rates, 1988–1999 (percentage of MTDC)

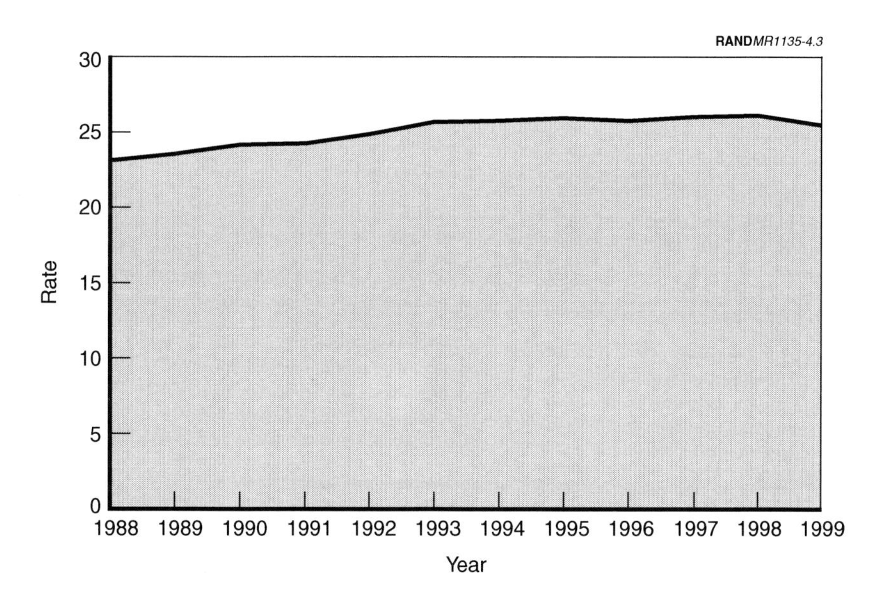

SOURCE: Database compiled from ONR and DHHS, 1999.

Figure 4.3—Trends in Facilities Component of Negotiated F&A Rates, 1988–1999 (percentage of MTDC)

Public universities consistently have lower average negotiated F&A rates than private universities do. Public universities in the past may have had less incentive to negotiate strongly for F&A reimbursement because much of their F&A costs were paid by state appropriations and some state governments did not give universities strong incentives to seek reimbursement of those costs from the federal government. In addition, there may be differences in the type of research performed at public universities or in the average age of their research facilities. Age and sophistication of research facilities both influence the cost structure in important ways. More sophisticated facilities involve higher construction costs as well as typically higher costs for operations, such as ventilation. Very old facilities are no longer depreciated or charged use allowances. In addition, universities may include some interest charges for financing facilities constructed since 1982, as discussed in Chapter Five. For these reasons, universities with newer facilities have more charges for facilities. Even newer facilities in public universities may use allowances rather than depreciation because their financial systems do not support accounting for depreciation. Private universities have been required to adopt depreciation accounting since 1988. Public universities will not have this requirement until 2001, as discussed in Chapter Five. Private universities' higher negotiated rates are likely explained by a combination of these factors: greater incentive to recover F&A costs, more new construction, and depreciation instead of use allowances.

The determination of the base over which to distribute F&A costs could be the most important factor in understanding rates. At some universities, the medical school is budgeted separately and is assigned an F&A rate separate from the rest of the university. Because medical schools engage in primarily biomedical research, which on average appears to use more specialized facilities and administration than other fields of science, they have higher average negotiated F&A rates compared with universities that lack medical schools in their base. Possibly, medical schools are simply better at documenting their costs, but the many special regulations applying to this field seem consistent with a higher cost structure in biomedicine. These regulations are described in Chapter Six and cover environmental health, animal care, and human subjects protection.

Figure 4.4 shows that public universities (represented by the two lower lines) have lower F&A reimbursement rates than private institutions do. This difference was about 13 percentage points in the late 1980s. The administrative cap, which took effect with the 1993 rates, affected private universities more than public, so this gap narrowed as private university administrative reimbursement rates declined. Recently, the gap between public and private rates has been about 10 percentage points.

The thinner lines in each section indicate the separate medical schools (private and public). Separate medical schools include those separately organized as well as those in institutions that negotiate distinct F&A rates for their medical schools as opposed to their other campus units. Although the data are ambiguous before 1993, since then medical schools in each group have shown somewhat higher average F&A reimbursement rates than all other institutions have. For this purpose, all other institutions include those that either do not have medical schools or do not negotiate separate F&A rates for medical schools.

Because all other institutions include a wide variety of institutions, we sought a closer comparison to examine possible differences in F&A reimbursement rates at medical colleges from other schools. Table 4.3 summarizes the results of this analysis. As shown in the table, the data had 27 or 28 medical colleges with rates in each year. For each medical college, we selected a specific comparison institution to build a comparison set. For institutions that had separate F&A rates for their medical schools and for the rest of campus, we used the rest of campus as the comparison. For stand-alone medical colleges (or where the main campus did not have a rate agreement), we selected a comparison institution within the same state and same type of control (public or private).[2]

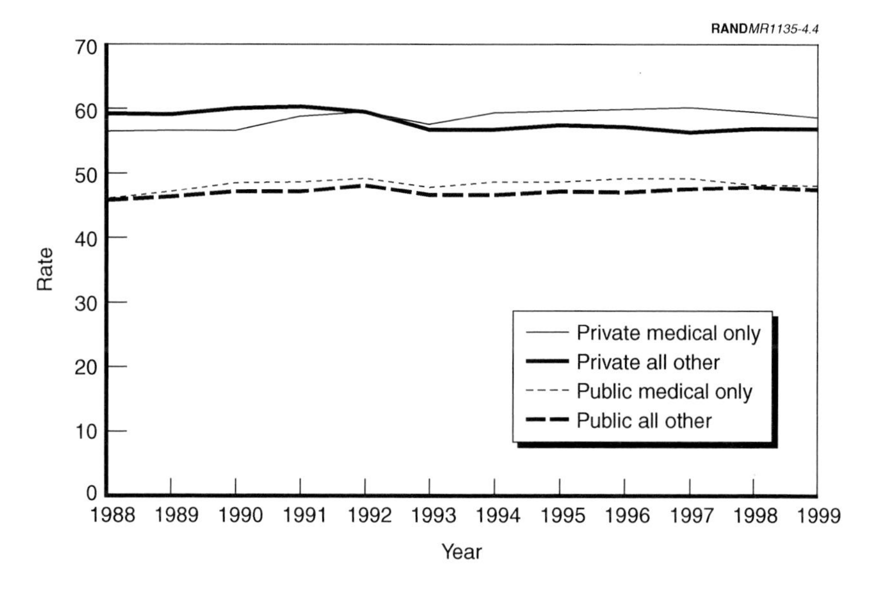

SOURCE: Database compiled from ONR and DHHS, 1999. Comparison institutions selected by RAND.

Figure 4.4—Trends in Public and Private (Medical Only and All Other) Negotiated F&A Rates, 1988–1999 (percentage of MTDC)

---

[2]For public university systems with separate medical colleges, the comparison institution would thus be a nonmedical member of the same system. In five cases where more than one comparison institution was eligible, we chose at random from the eligible choices. In two cases, no available comparison institutions were in the same state and type of control; for those, we selected an institution at random from a neighboring state with the same type of control.

The negotiated rates for administration in medical colleges and the comparison institutions are nearly identical year by year. But the negotiated rates for facilities diverge starting in the early 1990s. By the late 1990s, this difference has grown to about three percentage points.[3] Although many of the comparison institutions have no significant medical activity, some institutions with a single rate for the entire campus may include medical schools. Thus, these estimates of differences for medical colleges are likely to be slightly understated.

ESTIMATING ACTUAL RECOVERY RATES

Chapter Two showed that universities do not recover all of the negotiated F&A rate on every project. Schools differ in their recovery experience, depending on what their negotiated rate is. To illustrate this principle, consider that most USDA project grants are limited, by statute, to 14 percent of total costs for facilities and administration.[4] Theoretically, if a school's negotiated rate were low enough, it would see no reduction at all in its F&A reimbursement from these caps because its negotiated rate would yield reimbursement consistent with the

Table 4.3

Comparison of Negotiated F&A Rates at Medical Colleges and Comparison Institutions (percentage of MTDC)

	Medical Colleges				Comparison Institutions			
Fiscal Year	Facilities	Administrative	Total	Number	Facilities	Administrative	Total	Number
88	24.7	26.7	50.5	27	23.8	25.9	50.0	27
89	24.7	27.3	51.0	27	24.1	26.6	50.7	27
90	25.4	27.5	51.8	27	24.6	26.6	51.2	27
91	25.6	27.4	52.8	27	24.7	26.7	51.6	27
92	26.4	27.1	53.2	27	24.0	26.5	50.7	27
93	26.9	24.7	51.5	27	25.0	24.9	49.6	27
94	27.6	24.9	52.7	27	24.9	25.2	49.9	27
95	27.8	24.9	53.0	27	25.2	25.1	50.2	27
96	27.7	25.0	53.1	28	24.3	25.2	49.8	28
97	27.8	25.0	53.3	28	24.9	25.2	50.0	28
98	27.7	24.8	52.4	28	24.8	25.2	49.7	27
99	27.3	24.8	52.1	27	24.2	25.2	50.3	27

SOURCE: Database compiled from ONR and DHHS, 1999. Comparison institutions selected by RAND.

[3]Using a statistical t-test, we find that the difference in 1988 is insignificant. By 1999 the difference in facility rates is significant at the 0.10 level.

[4]Because the 14 percent limitation is calculated on total project costs, it can be roughly translated into an F&A rate of 20 to 30 percent of MTDC.

cap. For a school with a high negotiated rate, the cap greatly reduces F&A reimbursements on these projects. As a result of caps like this in several agencies, schools with high negotiated rates tend to recover less of their negotiated rates compared with schools with low negotiated rates.

To show the actual recovery by universities compared with their negotiated rates, we need information on universities' negotiated rates, their federal MTDC base, and their actual F&A recovery. Because expenditure data generally are maintained by each program within each agency (and some programs may not currently track F&A reimbursement separately), we do not have comprehensive government data to address this question.

These quantities are surveyed annually in the voluntary survey conducted by the COGR discussed in Chapter Two. About 130 COGR member institutions respond to the survey each year, including 80 percent of the institutions tracked by DHHS and ONR included in the above analyses. Over the four-year period between 1994 and 1997, we obtained 351 complete records from this survey, an average of 88 institutions each year. Figure 4.5 summarizes these data, showing the relationship between negotiated F&A rate and actual recovery rate (without regard to year).[5]

The figure shows that no group of institutions on average recovers its full negotiated rate. For institutions with the lowest negotiated rates, the recovery rate is closer to the negotiated rate than it is for institutions with higher negotiated rates. The small sample size in general—and in particular in the last few bars on the right—means that we cannot claim validity for small differences reported among groups in the figure. The figure should be taken in a general sense to indicate that cost-sharing of facilities and administration is widespread and proportionally higher in universities with higher negotiated rates. Institutions with higher negotiated rates, on average, also share a larger fraction of F&A costs. Although they may recover their full negotiated rate on some grants and contracts, in many instances they do not. Institutions with negotiated rates between 40 and 60 percent (a range that accounts for three-quarters of these institutions) report receiving, on average, about 77 percent of their negotiated F&A reimbursement.

These results indicate that any proposed changes in negotiated F&A rates may have unanticipated effects, because the relationship between negotiated rates and actual reimbursement is mediated by several factors, including agency limitations on F&A reimbursement and cost-sharing. It is possible that reduc-

[5]Because there is very wide variation in the actual recovery amounts from year to year for a given institution, as well as across institutions, we used medians to indicate the basic relationships in the data.

ing negotiated F&A rates could also reduce university cost-sharing resulting in no change in federal outlays for F&A reimbursements.

To summarize, overall trends in negotiated F&A rates have been basically level. Rates for administration have declined somewhat, while rates for facilities have increased about equally. In terms of negotiated rates, the administrative cap narrowed differences between public and private universities, but the differences remain noticeable. We do not know the precise explanation for these differences. The explanation is likely to involve some combination of more-expensive facilities at private universities, depreciation accounting instead of use allowances, and greater incentives, on average, for private universities to recover F&A costs than for their state-supported counterparts.

Examining data on actual recovery of costs by universities shows that negotiated rates do not tell the full story. There are statutory limitations on F&A recovery for all institutions, but these limitations have a greater proportional

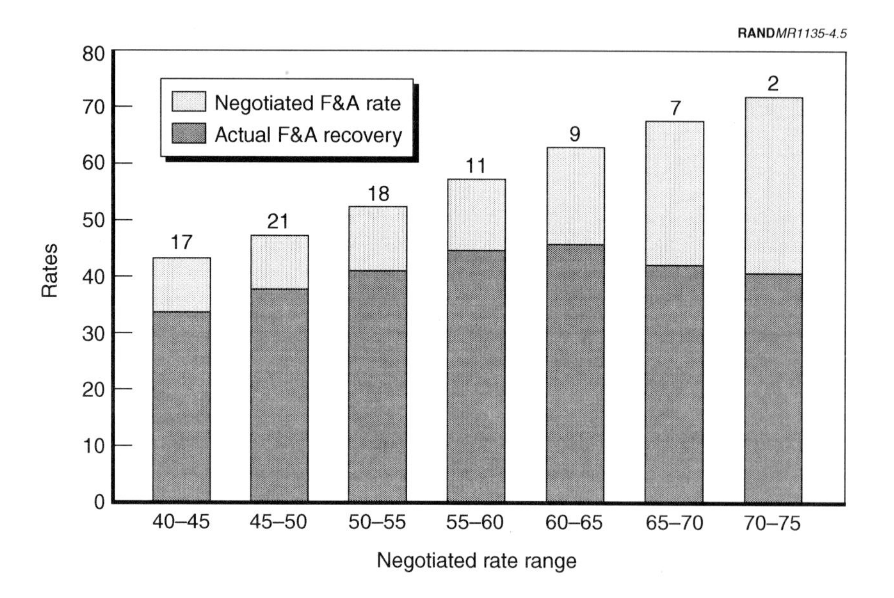

SOURCE: Database compiled from COGR, 1994–1997.

NOTE: Observations are grouped by negotiated F&A rate without regard to the year. The groups are formed with negotiated rates in the ranges 40–45, 45–50, 50–55, 55–60, 60–65, 65–70, and 70–75 percent. For each group the figure plots the median negotiated rate and the median actual recovery rate. The number of institutions in each group reported on the x-axis is computed by taking the number of observations in each range without regard to the year and dividing by four (the number of years of data) and rounded to the nearest whole number for display.

Figure 4.5—Comparison of Negotiated F&A Rates with Actual Recovery, 1994–1997 (percentage of MTDC)

effect when negotiated F&A rates are higher. For this reason and perhaps for other reasons, we find that universities with higher negotiated F&A rates also provide more cost-sharing for F&A costs. As a result, it is difficult to make judgments about federal outlays for F&A costs on the basis of data on rates alone or to predict the effect of changes in F&A reimbursement policies.

Chapter Five

ISSUE 3: THE IMPACT OF CHANGES IN CIRCULAR A-21

Appendix A contains a concise history of Circular A-21 and its revisions. From that history, we have identified recent changes that have had or may have impacts on negotiated facilities and administrative rates. These are the cap on administrative rates, library rates, infrastructure rates, the change in the equipment threshold, and introduction of the utility cost adjustment. The effects of these changes are summarized below:

- Cap on administrative rates: during 1993, the first full year it was in effect, the cap reduced negotiated administrative rates by approximately 2 percent; since then, administrative rates have remained constant.
- Library rates: these have remained constant since 1988.
- Infrastructure rates: these have increased gradually from nearly 6 percent in 1988 to approximately 9 percent in 1999, although some of the increase has been offset by reductions in operations and maintenance rates.
- Change in the equipment threshold: the magnitude of the effect of this change is not known, but it is expected to produce savings for both universities and the government because it reduces the need to track small items of equipment, especially personal computers.
- Utility cost adjustment: although the precise effect of this change is not yet known, it is unlikely to have a significant impact.

CAP ON ADMINISTRATIVE RATES

In 1991, OMB Circular A-21 was modified to impose a 26 percent cap on recovery of administrative costs in F&A rates. In the event that an institution decides not to compile the full documentation on all components of administration, it may claim a fixed allowance of 24 percent of modified total direct costs (or 95 percent of the most recently negotiated administration rate, if less). About eight institutions in the set we analyzed have rates for administration that equal 24 percent, although we do not know the precise number of institutions that use this provision. In these circumstances, the university does not have to prepare a cost proposal or document its costs for administration. (The institution must still justify facilities costs in the usual way.)

As a result of this change, most of the institutions we have data for are now charging exactly 26 percent for administration. In 1998, 93 out of 153 institutions had negotiated rates for administration of 26 percent. Most others have negotiated rates close to 26 percent. Figure 5.1 (which reproduces Figure 4.2) shows that the cap had a significant effect in 1993, when the provision took effect in most institutions, and that negotiated rates for administration since then have remained constant as a percentage of MTDC. The reduction in the rate appears to be about two percentage points on the negotiated rate.

Universities, especially private universities, may have taken the cap as a chance to reduce their administrative staffs in an attempt to live more within their reduced means. However, staff reductions also mean that individual scientists might now be doing some of the tasks formerly done centrally, especially if the need for the tasks has not vanished. Payments to individual scientists are considered direct costs of their projects and eligible for reimbursement. So the federal government might be paying more-expensive salaries for scientists to per-

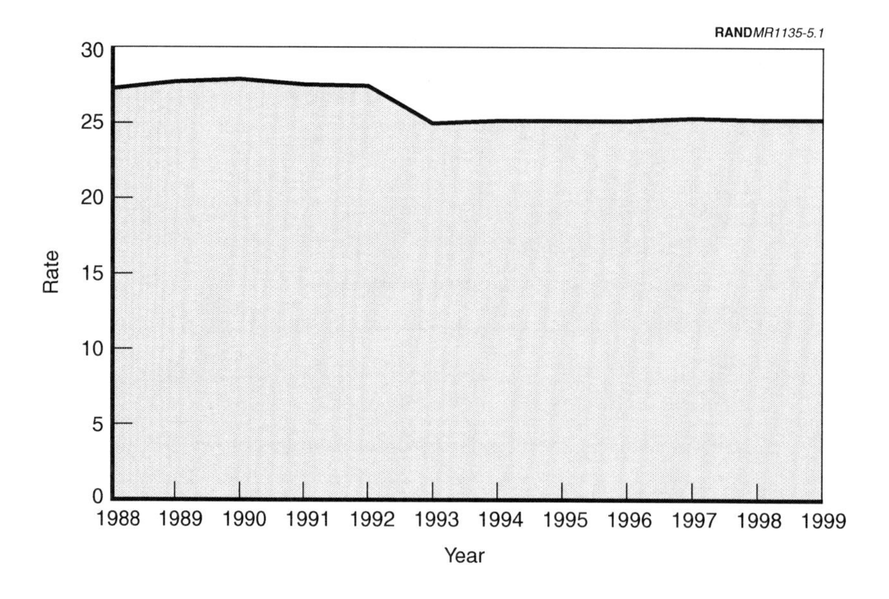

SOURCE: Database compiled from ONR and DHHS, 1999.

Figure 5.1—Trends in Administrative Component of Negotiated F&A Rates, 1988–1999 (percentage of MTDC)

form these tasks. We could not find hard evidence of these changes, and some debate has occurred concerning their existence and magnitude. However, there may have been some impact: administrative staffs have likely been cut, some workload eliminated, and some workload transferred to scientists.

The cap on administrative rates is certainly one method of reducing the government's expenses for university administration. But as Chapter Six describes in detail, many laws and regulations create administrative requirements for universities. Other ways of reducing the government's expenses for university administration might include finding ways to streamline the application of these laws and regulations that create administrative requirements. Savings from these changes could benefit both government and university.

LIBRARY RATES

From time to time, concerns have been expressed about the costs of libraries to support research. Universities have been allowed, in some cases, to conduct special studies to document who uses its libraries (undergraduate students, graduate students, faculty, and staff researchers). Based on these special studies, the university could apportion library costs to the university's research base. In 1996, it was proposed that special studies be eliminated for libraries, although this provision was rescinded in 1998. Figure 5.2 illustrates the data

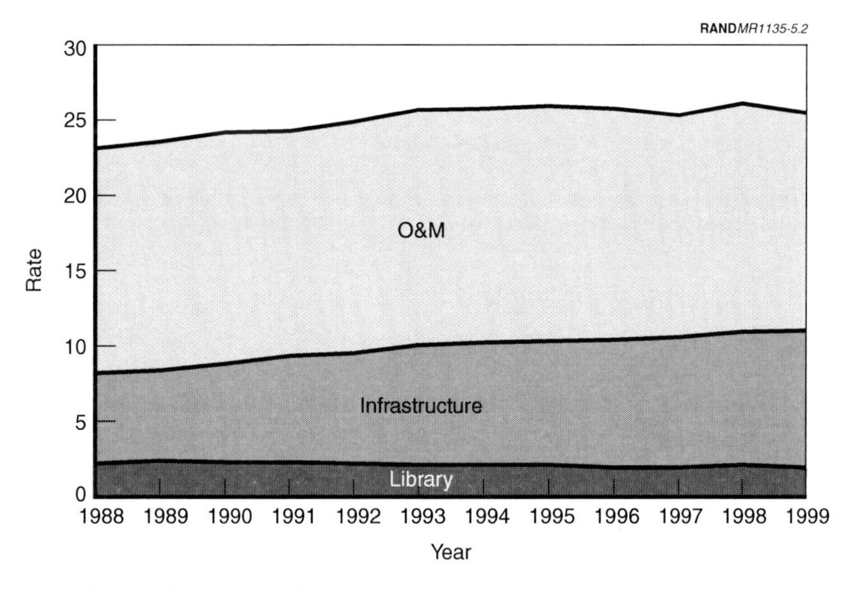

SOURCE: Database compiled from ONR and DHHS, 1999.

Figure 5.2—Trends in Facilities Components of Negotiated F&A Rates, 1988–1999 (percentage of MTDC)

from Table 4.2, showing that negotiated library rates have remained almost constant. There are concerns that the costs of operating libraries are rising, particularly the costs of maintaining access to current scientific journals. There have been no increases in negotiated rates for libraries, so either the MTDC base at universities has been rising enough to absorb increasing library costs or the universities are sharing these costs to prevent an increase in negotiated library rates.

INFRASTRUCTURE: INTEREST, DEPRECIATION, AND USE ALLOWANCES

Prior to 1982, universities could claim either depreciation or use allowances for buildings, equipment, and capital improvements used in research. Universities could thus recover the capital investment in facilities. But the costs of financing investments in facilities were not allowable under Circular A-21. That changed in 1982, when interest costs on debt used to finance buildings, equipment, and capital improvements for research became allowable in F&A rate negotiations.

This policy change has made it more attractive for universities to construct new facilities for research, because they can recover not only the depreciation or use allowance but also interest costs on the debt to finance the building. But there are several limitations on the recovery of these costs. First, the university must contribute a specified portion of the capital costs from its own funds in order to claim interest expenses (or else its interest expense is offset to make up for the required contribution). Second, Circular A-21 now requires a review of reasonableness and specified documentation for construction costs of major facilities. This review is designed to ensure that the costs of constructing facilities expected to support federally sponsored research are consistent with the planned research use.

Figure 5.3 shows that the effect of allowing interest to be included in rate negotiations became noticeable in the early 1990s, about 12 years after the changes in Circular A-21. It took time for universities to plan and execute new construction projects and for the interest costs of those projects to begin appearing in F&A rates. The increase in the amount of interest included in negotiated rates is consistent with a pattern of infrastructure modernization, encouraged by the 1982 change in Circular A-21.

The NSF Survey of Research Facilities gives us an indication of the changes in research facilities over the past decade. According to that survey, between 1988 and 1998, research space increased 28 percent (from 112 million to 143 million square feet). In addition, according to the survey respondents the portion of research space rated "suitable for the most scientifically sophisticated research" increased from 24 percent in 1988 to 39 percent in 1998 (NSF, 1998). So in terms of both quantity and quality, there has been a substantial modernization and expansion of university infrastructure.

It is likely that the effects of this modernization will continue to be apparent for many years. Bond debt is typically outstanding for 20 to 30 years, so we would expect to see the interest costs for a building project included in F&A rate negotiations over that period. Debt used to finance major equipment purchases would likely have a shorter payoff period, owing to the shorter useful life of equipment compared with buildings.

Although the amount of depreciation and use allowances in negotiated rates increased prior to 1994, this component did not increase during the period when the interest component rose. Overall, recent increases in building and equipment components have been offset by reductions in the operations and maintenance component. We are not able to observe actual cost experience, only the results of rate negotiations. One explanation for these changes is that newer buildings and equipment are more energy efficient and easier to maintain, leading to lower utility expenses and maintenance bills. Based on Table 4.2, between 1993 and 1999 infrastructure components in negotiated F&A rates increased by about one percentage point whereas the operations and maintenance component decreased by about one percentage point, leaving overall facilities rates unchanged.

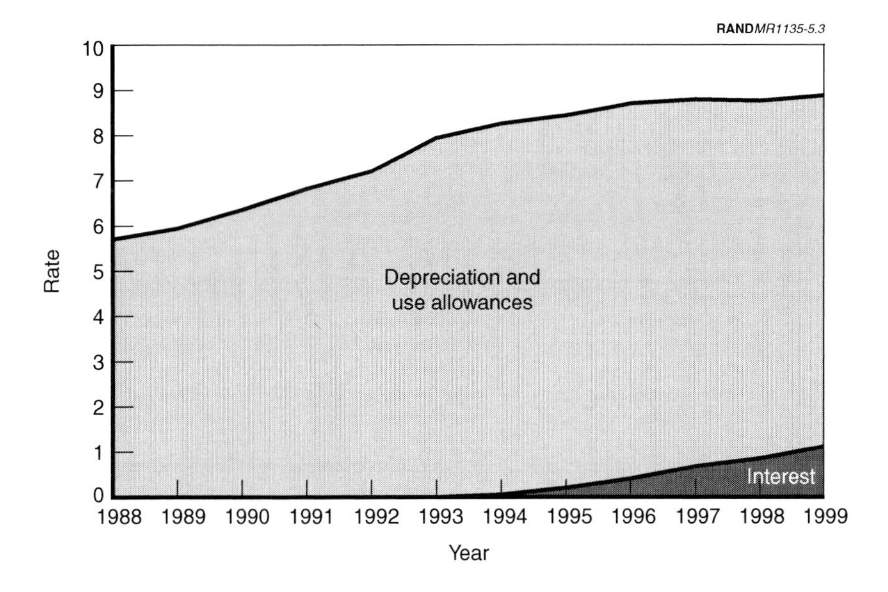

NOTE: Based on DHHS institutions only.
SOURCE: Database compiled from ONR and DHHS, 1999.

Figure 5.3—Trends in Infrastructure Components of Negotiated F&A Rates, 1988–1999 (percentage of MTDC)

Today, many public institutions rely on use allowances rather than depreciation. Although it is not a change in Circular A-21, beginning in 2001 public institutions must adopt asset accounting methods, leading to much wider use of depreciation, rather than use allowances, for public institutions. The effects of this change are unknown but could lead to increased recovery of infrastructure costs by public institutions. Private institutions have been required to adopt asset accounting since 1988.

In summary, Figure 5.3 shows that the infrastructure component in negotiated F&A rates has increased since the late 1980s. The direct effect of allowing interest charges is noticeable in rates, but does not account for all of the increase in the infrastructure component. Allowing interest costs to be included in rate negotiations may have stimulated construction, which in turn may have led to greater negotiated rates for depreciation or use allowances.

As shown in Figure 5.2, recent increases in negotiated rates for the infrastructure component have been offset by reductions in the operations and maintenance component. Despite the lack of change in overall facilities rates recently, changes in Circular A-21 to allow interest costs appear to be encouraging institutions to modernize their research facilities.

CHANGE IN EQUIPMENT THRESHOLD

Equipment is formally defined as personal property having a useful life of more than one year and costing more than a set threshold to acquire. Institutions are required by Circular A-110 to conduct a physical inventory at least every two years to verify that each asset is usable and needed for its purpose. Prior to 1996, the equipment threshold was $500. In 1996, Circular A-21 allowed institutions to raise the threshold to the institution's own internal threshold used for accounting purposes, up to a maximum of $5,000. Universities are not necessarily free to raise their own equipment threshold. For example, state universities are generally governed by state rules that specify the threshold for equipment. These thresholds are significant because of the effort involved in tracking equipment. For all equipment, the institution must maintain accounting records to track the purchase and use of the item and to record either the depreciation or the use allowance applicable to each item over its life.

The increase in the threshold has many benefits for institutions. Some institutions estimate that they can eliminate capital asset tracking on 80 percent or more of the items they were required to track with the $500 threshold. This should generate administrative savings because institutions will not have to maintain accounting records over the life of each asset and conduct inventories of the affected assets. The government should experience direct savings because of reduced auditing costs and possible savings if university administrative savings are passed through in lower negotiated administrative rates.

UTILITY COST ADJUSTMENT

Utility costs for research facilities are often higher than those for other campus buildings because research space often houses equipment that has higher demands for electricity, heating, air conditioning, and ventilation. Before June 1, 1998, many universities conducted special engineering studies to determine the amount of utility costs allocable to research facilities. Instead of these studies, which were characterized as expensive to perform and complex to validate, the revisions to Circular A-21 permit the 66 universities that were granted additional utility recovery on the basis of special studies to claim a fixed adjustment of 1.3 percent. Special studies were eliminated. Beginning in 2002, institutions not on the original list may ask to have this 1.3 percent provision applied to them. OMB and the cognizant agencies will have to develop criteria to determine which institutions will be eligible and perhaps adjust the allowance.

The 1.3 percent adjustment is based on the average amount allowed under previous special studies for utilities. It is intended to have no net effect on F&A reimbursements. But individual universities might fare better or worse under this system. If a university experiences a substantial increase in research volume, it receives the 1.3 percent utility adjustment applied to its total MTDC base. In this case, the university might increase its F&A recovery. On the other hand, some universities may see reduced recovery. A university able to document higher utility costs in research space than average for this group can no longer rely on special engineering studies and must accept the 1.3 percent of MTDC amount, even though it is lower than its previous negotiated rate. Because this provision is new, the precise effect on average rates and recovery is unknown. Because the adjustment was set at the average level justified by the old studies, the utility cost adjustment is unlikely to have a noticeable effect on average.

Chapter Six

ISSUE 4: THE IMPACT OF FEDERAL AND STATE LAW ON F&A RATES

Federal, state, and local laws and regulations entail significant costs for universities engaged in federally sponsored research. While the laws and regulations are enacted for worthwhile purposes, they bring about real costs. Universities must design and maintain their facilities to comply with these requirements, and they must create mechanisms to ensure and certify compliance. The costs are substantial in both facilities and administration. The imposition of a cap on administrative costs that can be included in F&A rate negotiations means that universities must bear all administrative costs over the cap. One option for reducing administrative costs would include reducing some of the administrative requirements generated by these laws and regulations. The potential savings could benefit both the universities and the government.

THE SPECTRUM OF LAWS AND REGULATIONS AFFECTING RESEARCH UNIVERSITIES

Facilities and administrative costs at colleges and universities are governed by the principles in Circular A-21. Circular A-21 specifies which costs are allowable for purposes of negotiating F&A reimbursement rates. But many other laws and regulations—at the federal, state, and local level—affect costs for facilities and administration. These laws and regulations support a number of objectives, including the desire to protect the health and safety of humans and animals and to promote good stewardship for federal research funding. In response, research universities have created administrative departments and panels to monitor compliance efforts in many areas of their operations: hazardous waste storage and disposal, occupational safety and health, animal care, and human subjects protection. The cost of compliance includes inventories; certification programs; training programs for faculty, staff, and students; and legal expenses. These laws and regulations can also incur facilities costs. Facilities must be constructed, renovated, and operated in accordance with federal, state, and local laws.

The nature of these laws and regulations has changed markedly over time. In the past, it was often presumed that laboratory scientists were in the best position to understand and manage the risks of the chemicals and biological agents they used. Other areas of concern were also monitored at the laboratory or facility level. But over time, increasingly sophisticated regulations have required new specialized personnel. One university explained:

> [W]hile many of these standards began as ancillary requirements that could be supervised by facilities employees, they have rapidly grown into jobs requiring specialists to manage the complex and sometimes bureaucratic laws and regulations that they represent. (Goldman, 1999.)

The costs of these compliance efforts directly affect the components of F&A rates. Costs for compliance affect both facilities and administrative components. Administrative costs generally are recurring costs to staff oversight committees. Facilities costs may be onetime costs involved in construction of new facilities or major renovations of older facilities, but they may also be ongoing in character. We have seen no systematic data on the costs of these requirements. We obtained some detailed information from one public university whose research base was slightly above the average for the COGR survey institutions. Although we are not sure how this information would generalize, the university maintains a wide range of research programs and does not appear to represent an extreme case. The university reported that compliance with facilities requirements necessitates so many improvement projects for existing facilities that it is infeasible to undertake them all at once. This university has committed $1.2 million per year, indefinitely, for facilities improvements to enhance compliance with hazardous waste, occupational safety, animal care, and other facilities regulations (Goldman, 1999). This annual expenditure represents roughly 2 percent of the university's MTDC base (which would correspond to two percentage points of the university's F&A rate). This is only a partial estimate of the costs of compliance. This estimate does not include the costs of compliance associated with major building renewal or new construction projects that the university undertakes. Neither does it include the costs of administrative oversight each year to track compliance, train people, and make reports. The additional costs as part of facilities construction, operation, and administration would increase this estimate of compliance costs substantially.

As stated above, the university's experience is difficult to generalize, but the range and complexity of requirements discussed in the following sections point toward a significant influence on both facilities and administrative costs.

HAZARDOUS WASTE DISPOSAL AND LABELING

Prior to the 1980s, research scientists working with various chemicals were presumed to know how to handle and dispose of them. The 1980s marked an evolution in how to handle, store, use, and dispose of hazardous materials. Over the past 19 years, safety information, standard procedures, regulations, and laws have been developed addressing the acquisition, storage, and use of hazardous chemicals.

Although we did not find precise estimates, research universities appear to produce a very small amount of the nation's hazardous waste.[1] But research institutions must follow federal, state, and local guidelines just as industrial firms do. Industrial operations, though, often produce large quantities of a few waste products as a result of a specific manufacturing process. In contrast to the industrial firms, research universities produce "hundreds—or even thousands—of different waste products, some in quantities as small as a gram." (Andrews, 1991.)

There are many federal laws and regulations concerning how, when, and where to store and dispose of hazardous materials. Research universities must comply with federal laws, including the Resource Conservation and Recovery Act (RCRA), the Comprehensive Environmental Response, Compensation, and Liability Act (CERCLA, also known as "Superfund"), Clean Air Act, Clean Water Act, Toxic Substances Control Act (TSCA) and the Hazardous Materials Transportation Act (HMTA).[2]

States may further regulate hazardous materials. In California, Proposition 65 requires informing the public of environmental exposures to chemicals above specific risk levels. This law requires the posting of warning signs in or near areas that contain any of the cancer-causing agents or reproductive toxins known to the state to cause cancer or reproductive harm. As of 1999, the state had listed 660 substances as subject to this act. Businesses, including higher education institutions, must notify employees and the public if they may be exposed to one or more of these substances, unless the business chooses to evaluate the specific risk posed and demonstrate that it is below legally specified thresholds.

California also regulates acutely hazardous materials (AHM) with local implementation subject to limited oversight by the state's Office of Emergency Services. When the maximum quantity of AHM in one of its buildings exceeds

[1]Andrews (1991) claims that the fraction of hazardous wastes produced by academic laboratories is "probably less than 1 percent" of the national total.

[2]Title 49 of the Code of Federal Regulations (C.F.R.).

specified threshold quantities, a university, like other businesses, must register as an AHM handler.

The local government could require the university to prepare a Risk Management and Prevention Program (RMPP) for one or more on-site acutely hazardous materials. If required by the local government, the RMPP would specify risk-reducing changes in equipment, operations, and maintenance. The RMPP involves engineering and operational reviews of the AHM-handling system, evaluations of the off-site consequences of likely AHM accidents, and the implementation of on-site improvements to reduce the likelihood and severity of any AHM accidents. Release of AHM must be reported to local administering agencies. In addition, the federal Environmental Protection Agency has established for each AHM a reportable quantity under Superfund, such that any unauthorized release of an amount exceeding the reportable quantity also triggers immediate reporting requirements.

Local ordinances may require universities to comply with a variety of standards, such as those listed here.

- The Uniform Fire Code and Uniform Building Code, adopted in many jurisdictions.
- Permits for facilities that store hazardous materials. A research university must submit a plan for monitoring stored materials to detect releases; for posting emergency procedures where hazardous materials are stored; for regular testing and inspection; and for regular maintenance, repair, and replacement of storage facilities and emergency equipment.
- Inventory statements. Universities must submit an inventory statement (hazardous materials inventory statement) as part of the hazardous materials management plans for each building containing such materials.[3] Failure to submit an adequate inventory exposes the university and individual researchers to criminal and civil penalties.
- Toxic gas regulations. An ordinance may require seismic protection, security, leak testing, separation of incompatibles, protective plugs, emergency drills, fire extinguishing systems, and annual maintenance applied to all regulated gases. Specific control mechanisms and procedures for different classes of gases may also be required.

[3]The statement list must contain general chemical names; common/trade names; major constituents for mixtures; manufacturers; United Nations or North America shipping numbers, if available; aggregate quantity ranges; and carcinogen identification forms. The inventory is a public record.

To take one example, as a result of state and federal environmental regulations, local authorities have increased requirements for the enforcement of lower mercury discharge. To avoid abatement orders from local authorities, universities must increase mercury discharge-monitoring and prevention measures for such research fields as dentistry, chemistry, and pharmacy.

To ensure compliance with federal, state, and local environmental laws and regulations, universities establish environmental health and safety departments. These departments develop the university's hazardous materials management plan. Regulations require the plan to contain a detailed floorplan that shows all hazardous materials, including wastes, the hazard class, and the quantity range for each class aggregated within each storage facility. The plan must also include a description of the methods used to ensure separation and protection of stored hazardous materials from factors that may cause fire or explosion, monitoring methods, emergency procedures, maintenance schedules, and record-keeping forms.

In addition to the environmental health and safety departments, universities generally form oversight panels that report to the president through a university official, such as the provost or dean of research. These administrative panels monitor the university's research activities by reviewing proposals that may involve hazardous materials, including chemical, biological, and radiological agents. The university official responsible for oversight of university research will frequently undertake to produce a research policy handbook describing the responsibilities of principal investigators and their staffs to be in compliance with federal, state, and local laws and regulations.

Specialist environmental managers are needed for local and EPA permitting, preparation and revision of hazardous materials plans, abatement of hazards, and developing processes that will keep the institution and its personnel current with environmental regulations.

The universities' response to federal, state, and local regulation entails significant costs. When building new facilities and renovating old ones, the cost of construction includes satisfying environmental standards based on law and regulation and obtaining the proper permits from local authorities. Operations and maintenance costs include ongoing utility costs for required ventilation, personnel and materials to maintain storage tanks, and complying with abatement plans. Administrative costs include staffing for the environmental health and safety department, maintaining databases of materials on campus, production of handbooks for researchers, and training.

OCCUPATIONAL HEALTH AND SAFETY

The federal Occupational Safety and Health Administration requires following OSHA Laboratory Standards and protecting workers against potential exposures to HIV, hepatitis B virus, and other bloodborne agents by maintaining and mandating the establishment of a written Exposure Control Plan. The Exposure Control Plan requires that the university

- identify the tasks and procedures where occupational exposure to bloodborne pathogens is likely to occur; and
- specify a schedule for implementing record-keeping; housekeeping; communicating hazard information; following Universal Precautions; implementing engineering and work practice controls; ensuring medical follow-up for workers who have an exposure incident; and providing training, personal protections equipment, and hepatitis B vaccinations, including a training course prior to vaccination.

The OSHA Laboratory Standard is tailored for individual laboratories and exempts research and clinical labs from state requirements intended for industrial environments. A "laboratory scale" chemical is defined as one that is not part of a production process and can be manipulated by one person. The university is required to generate a comprehensive Chemical Hygiene Plan that protects workers' (including students') health around chemical hazards. The plan includes control measures, equipment performance measures, and the implementation of written standard operating procedures.

State Occupational Safety and Health laws may further regulate carcinogens and their use. State OSHA laws may designate areas where the use of certain carcinogens is permitted and require that any spills, leaks, and possible exposures be reported to the university's department of environmental health and safety immediately. The university's department of environmental health and safety may in turn be required to report the exposures to the state within 24 hours. Failure to report in the appropriate time frame could lead to a serious citation and fine by the state. Federal and state OSHA regulations mandate training, the use of personal protective equipment, standard operating procedures, labeling, emergency measures, and posting.

To maintain compliance with occupational health and safety standards, universities incur costs to bring facilities up to standard and for administrative organizations to document and monitor compliance.

ANIMAL CARE

Research universities are required to comply with three major federal laws and regulations when working with animals: the Animal Welfare Act (AWA), the Public Health Service Policy on Humane Care and Use of Laboratory Animals, and the National Academy of Sciences *Guide for the Care and Use of Laboratory Animals.*[4]

Originally enacted in 1966, the legislative intent of the Animal Welfare Act (AWA) was to ensure that pets or animals in research facilities or for exhibition purposes are provided humane care and treatment and ensure the humane treatment of animals during transportation in commerce.[5]

The AWA establishes the minimum requirements for the care of animals, including their housing, feeding, shelter, and exercise (AWA, Section 13(a)(2)(A)). Research universities are required to meet these standards:

- Show upon inspection, and to report at least annually, that the provisions of this Act are being followed and that professionally acceptable standards governing the care, treatment, and use of animals are being followed by the research facility during the actual research or experimentation (AWA, Section 13(a)(7)(A)).
- Establish at least one Institutional Animal Care and Use Committee (IACUC or other similarly named committee) consisting of three members to inspect at least semiannually all animal study areas and animal facilities of such research facility and review as part of the inspection the practices involving pain and the condition of the animals to ensure compliance (AWA, Section 13(b)(1)).
- Provide for the training of scientists, animal technicians, and other personnel involved with animal care and treatment in such facility (AWA, Section 13(d)).

---

[4]Other laws and regulations setting requirements for animal care include EPA, Good Laboratory Practice Standards (40 C.F.R. Pt. 792 (1997)); FDA, Approved Animal Drug Products; FDA, Good Laboratory Practice for Nonclinical Laboratory Studies (21 C.F.R. Pt. 58 (1997); FDA, Good Laboratory Practice Regulations; Minor Amendment: Toe Clipping (54 Fed. Reg. 1989)); NIH Intramural Guidelines for the Euthanasia of Mouse and Rat Fetuses and Neonates (February 1997); NASA Principles for the Ethical Care and Use of Animals (1996); Public Health Service Policy on Humane Care and Use of Laboratory Animals (1986, reprinted 1996); Public Law 99-158—November 20, 1985, Health Research Extension Act of 1985, Animals in Research; Public Law 102-346—August 26, 1992, Animal Enterprise Protection Act of 1992; USDA, Agricultural Research Service Directive 635.1, Humane Animal Care and Use (1990); USDA, Horse Protection Act; and USDA, Horse Protection Act Regulations.

[5]Originally titled the Laboratory Animal Welfare Act, Amended 1970, 1976, and 1985, the Animal Welfare Act (7 U.S.C. 2131–2156).

The Office for Protection from Research Risks of the DHHS develops, implements, and oversees compliance with the Policy on Humane Care and Use of Laboratory Animals.[6] For any research university participating in the sponsored projects of the DHHS Public Health Service (PHS), documentation must be provided of compliance with the Policy on Humane Care and Use of Laboratory Animals through a Letter of Assurance. The essential elements of this letter must describe the following:

- The institutional program for care and use of animals, including information about the research university's Institutional Animal Care and Use Committee; the employee health program for those in frequent contact with the research animals; and the gross square footage, average daily census, and annual use of each animal facility.
- The research university's institutional status as either accredited by the American Association for the Accreditation of Laboratory Animal Care or nonaccredited. If an institution is not accredited it must establish a plan, including specific guidelines for correcting any departures from the recommendations in the National Academy of Science's *Guide for the Care and Use of Laboratory Animals.*
- The Institutional Animal Care and Use Committee. In contrast to the AWA, the PHS policy requires a minimum of five members.

The *Guide for the Care and Use of Laboratory Animal*s provides guidelines to research universities on animal care programs and facilities. The NIH and most federal funding agencies require its use in determining the appropriate standards for animal care.

USDA's legislative mandate on animal care and use is somewhat broader than DHHS's, covering exhibitors and dealers in addition to research use. Regulatory approaches also differ. For example, USDA regulates specifically covered species whereas the DHHS policy applies to animal research conducted at facilities receiving DHHS funding. The two departments cooperate on the regulation of animal care and use.

Research universities must either maintain or modernize their facilities to be in compliance with these standards. Modernization of facilities can entail significant expense in construction. One social science laboratory alone reported that the cost of remodeling animal quarters was more than $3.5 million. In addition to construction costs, specific environmental controls and backup systems must be maintained, including virus-isolation mechanisms. Research universi-

---

[6]The 1985 Health Research Extension Act requires all medical research funded through the NIH to conform with the policy.

ties report that the operations and maintenance costs, such as ventilation, are significant to comply with animal welfare regulations.

Universities thus incur administrative costs for the review panels required by laws and regulations, as well as facilities costs to build, renovate, and operate facilities in compliance.

HUMAN SUBJECTS

Congress has authorized the DHHS to regulate the protection of human subjects when they are part of a research project.[7] Specifically, the law applies to any entity that receives federal funding for projects in behavioral or biomedical research on human beings. The law requires such entities to establish an institutional review board to review all research in these areas with the institution. It is the responsibility of DHHS to promulgate standards for review, composition of the review boards, and support of the review boards by the institution. The standards are based on ethical research policies developed at each institution based on the federal regulations. Individual researchers, the institutional review board, and the senior university administrators all play a role in ensuring compliance with public policy and research ethics.

The institutional review board must consist of at least five members, typically faculty but including at least one member of the community who is unaffiliated with the institution. The board, which is often larger than the minimum, must represent a range of professions and perspectives. The members cannot all come from the same field.

The regulations require review boards to maintain records of their evaluations of each research proposal. The regulations also set standards for the informed consent of research subjects and establish guidelines for projects that do not require a full review. Even in those cases, the review board must still maintain records for all projects concerned with human subjects.

In the regulations, special requirements for scrutiny apply to potentially vulnerable populations, including fetuses, pregnant women, human ova, prisoners, children, and the cognitively impaired. In particular, the review board must give consideration to whether its members are adequately knowledgeable about the specific vulnerable population. If not, they may seek outside expert advice to aid the board. For Department of Education research that concerns either handicapped children or mentally disabled people, the review board must have a person primarily concerned with the welfare of the relevant subject

---

[7]42 U.S.C. 289 and Protection of Human Subjects (45 C.F.R . 46).

population.[8] In all cases where research involves subjects from a potentially vulnerable population, there are specific standards required of researchers for each category of vulnerable population. In some cases, the review board may need to authorize regular monitoring of the research to guard against unanticipated risks to the subjects.

A number of expenses are associated with compliance with human subject regulations. These are usually categorized as allowable administrative costs. Universities are required to provide space and support staff for meetings and record-keeping of the boards. There may be costs for the members of the boards, as well. Because a large university can generate many research proposals requiring human subjects review, some universities indicate that they fund substantial release time for faculty who chair or serve on these review boards. This release time may be classified as an administrative expense. If the review board determines that it needs outside evaluations of research proposals, or ongoing monitoring of projects with vulnerable populations, the university may incur additional expenses for these services.

FOIA DISCLOSURE OF RESEARCH DATA

Pursuant to the requirements of Public Law 105-277 (The Omnibus Consolidated and Emergency Supplemental Appropriations Act, 1999) OMB has implemented requirements for research institutions to release data that are developed with federal research grants through the Freedom of Information Act (FOIA). These regulations were added to OMB Circular A-110 on September 30, 1999. The requirement applies to research data that are used by a federal agency "in developing an agency action that has the force and effect of law."

The circular authorizes agencies to collect "a reasonable fee equaling the full incremental cost of obtaining the research data. This fee should reflect costs incurred by the agency, the recipient, and applicable subrecipients." Because this is a new requirement, there may be some unanticipated effects. If the fees do not cover the costs of determining which data can be released and in what form, the universities may bear some of these costs. In addition, universities may have to establish additional administrative offices to receive and coordinate FOIA requests. If these unreimbursed costs prove to be substantial in practice, they would tend to increase expenditures for administration.

---

[8]34 C.F.R. Pts. 350 and 356.

COST ACCOUNTING STANDARDS

All universities are required to comply with the four Cost Accounting Standards (CAS) for educational institutions, as set forth in OMB Circular A-21. In addition, all universities that receive aggregate sponsored agreements totaling $25 million or more during their most recently completed fiscal year are required to file a Disclosure Statement, Form DS-2. The purpose of the DS-2 is to document the institution's specific set of cost accounting practices in enough detail to document that its procedures meet the requirements of the CAS.

The four cost accounting standards for universities require that universities implement accounting procedures that ensure (1) consistency in estimating and reporting costs across departments, programs, and functions; (2) consistency in allocating costs to direct or indirect pools; (3) identifying unallowable costs; and (4) using the same fiscal year for all programs and functions within the institution. Many universities have used somewhat different methods to account for costs in different areas of the university and therefore have accounting systems that do not always conform to the new CAS requirement. Changing complex accounting systems and methods of accounting as required for compliance with CAS entails significant onetime costs.

A recent survey of 18 major research universities by COGR reported that the average university in this group spent $200,000 to prepare its initial DS-2 disclosure statement, frequently involving outside consultants. Depending on the size of the university's MTDC base, this would represent between approximately 0.5 and 1.0 percent of MTDC, or 0.5 to 1.0 percentage points of the F&A rate during the first year.

Ordinarily, the costs of changing accounting methods and systems and preparing required disclosure statements are allowable administrative costs. However, the CAS Disclosure Statement requirement applies to the largest universities. Because larger universities are more likely to have higher negotiated rates for administration, many universities subject to the disclosure statement are already limited by the 26 percent administrative cap and hence cannot increase recovery to recoup these costs. Ongoing costs are likely to be much lower, because most of the costs are a result of initial changes and filings.

To summarize, the requirements of CAS are causing large universities to revamp their accounting systems and procedures. If it were not for the administrative cap, these changes would result in one-year increases in administrative rates of between 0.5 and 1.0 percent. As a result of the caps, it appears that most of these costs will be borne entirely by the universities. We expect that these costs are primarily onetime and ongoing costs should be much smaller.

CERTIFICATIONS AND ASSURANCES

To comply with federal requirements concerning health and safety, the environment, animal care, human subjects, and fiscal accountability, universities undertake the measures described in the preceding sections. In addition, universities are required to adhere to other major federal policies. Two of these requirements are that universities uphold federal nondiscrimination policies and maintain a drug-free workplace in accordance with relevant laws. For all of these areas of law and regulation, universities must provide certifications and assurances that they are in compliance.[9]

Certifications and assurances are administrative activities. The university's sponsored programs office coordinates the preparation of the required certifications and assurances, some of which can be submitted one time to cover a whole year. Others must be submitted with each award application. The costs of preparing these documents and packaging them with each application are part of the administration component in F&A rates. There may be opportunities to consolidate the processing of certifications and assurances, because these requirements generally are implemented institutionwide, rather than as part of each grant or contract proposal. Consolidation would reduce administrative costs for both universities and the federal government. In addition, reductions in these requirements might be passed through in lower proposed administrative rates.

STATUTORY LIMITS ON F&A REIMBURSEMENT

Certain agencies and programs do not reimburse universities for the full cost of projects, including direct costs and the negotiated F&A rates. The NSF requires grantees to share at least 1 percent of project cost, including direct and F&A costs. In practice, cost-sharing has been higher than the minimum in some NSF programs. NSF grants typically do not include support for faculty effort during the academic year, meaning that the university must bear those costs (including the corresponding share of F&A costs) from other sources. Other agencies also have cost-sharing policies that apply to direct project costs. For example, NIH has a legislatively imposed salary cap for research grant participants, currently $136,700 per year. If any participant in a research project earns a salary higher than the cap, the institution must pay all costs over the cap, including the corresponding share of F&A costs.

Statutory limits also specifically apply to F&A costs in three agencies. These limits apply to all USDA competitive research grants, NIH grants for predoctoral

---

[9]See NSTC (1999), Chapter 5.

and postdoctoral training, and certain Department of Education grants. As discussed in Chapter Two, agricultural extension programs are generally characterized as the federal share paying only for direct costs and the state share paying for all F&A costs as well as some direct costs.

NIH awards under the National Research Service Act of 1974 provide institutional grants for predoctoral and postdoctoral training, as well as individual grants. Under DHHS policy, institutional grants are limited to an F&A cost reimbursement of 8 percent of MTDC. Individual grants include a fixed allowance paid to the fellow's institution rather than F&A reimbursement. The allowance covers such expenses for the individual fellow as research supplies, equipment, travel to scientific meetings, and health insurance. If those allowed expenses do not exhaust the 8 percent of MTDC for the grant, any remainder may be applied to the institution's administrative costs.

NIH grants for career support to faculty—as opposed to project funding—also come with an 8 percent F&A cost reimbursement limit. Other agencies that award institutional training grants, including USDA and the Department of Education, follow the same 8 percent F&A cost reimbursement formula.

USDA operates under several congressional limitations on F&A reimbursement. The National Research Initiative specifies that no more than 19 percent of an award can be used for F&A costs. The appropriating language, however, further reduces that to 14 percent of an award. Certain programs maintain the 19 percent level. Two USDA programs allow recipients to claim their full negotiated rate: some SBIR awards to small businesses and higher education awards to tribal colleges. USDA believes that small businesses would find it too difficult to participate in its grant programs if they could not recover their full indirect costs. Universities, on the other hand, are presumed able to share F&A costs from other sources.

USDA calculates reimbursement using a college's or university's negotiated F&A rate against the MTDC for a project. If that amount is below 14 percent of the total project cost (or 19 percent in certain programs), then the award is made for the *lower* amount of F&A costs.

The largest part of USDA's awards are formula funding for research and extension that require matching by states or institutions and are construed to contain no reimbursement for F&A costs. For these awards, the recipient institution or its state must share at least half of the total costs, including absorbing all facilities and administrative costs.

In Chapter Two, we estimated that between $0.7 and $1.5 billion of the negotiated F&A reimbursement rates are not actually reimbursed to universities each year. In addition, universities must support costs for personnel who do not charge their time directly to projects, such as faculty time during the academic year and the costs of researchers for salaries above the NIH salary cap. The costs not borne by federal agencies must be borne by others. In the case of state-supported universities, many of these costs are likely to be borne by state appropriations, which cover faculty salaries, building operations, and maintenance. At private universities, the situation is more ambiguous. Private universities receive funds from students and their families for tuition and from private donors for many purposes. We conclude that some funds from these sources cover unreimbursed costs of faculty time in research and F&A costs not provided by research sponsors, including the federal government, industry, and foundations.

Chapter Seven

ISSUE 5: OPTIONS TO REDUCE OR CONTROL THE RATE OF GROWTH OF FEDERAL F&A REIMBURSEMENT RATES

Chapter One presented data that show universities share substantially in the costs of research supported by the federal government. The federal government awards about $15 billion to higher education institutions for research. The institutions provide about $5 billion more in funds, some of which pays for facilities and administrative costs on federal awards. Chapter Three reviewed available data on total costs and concluded that facilities and administrative costs represent about 31 percent of total research costs at universities. Chapter Two used two methods to estimate the amount of F&A costs actually paid by the federal government. This share is between 24 and 28 percent of federal outlays, lower than the 31 percent of total costs that these true costs represent. Based on the difference in these figures, we concluded that universities are sharing F&A costs. Universities might share even more of these costs if federal support were reduced. We do not know the mechanisms universities use now to share these costs or those they might use if federal support were reduced. These mechanisms could include supporting some research costs with other funds, such as state appropriations, private gifts, endowments, or tuition revenue.

Chapter Four showed that overall negotiated F&A reimbursement rates remained constant over the 1988–1999 period. Although we do not have comprehensive data on actual reimbursements for F&A *costs*, available evidence shows that *negotiated rates* have remained steady. Therefore it does not seem likely that actual reimbursements are increasing as a percentage of research awards.

There is substantial variation in agency experience in paying for F&A costs. Some agencies appear to operate successful programs with caps on F&A costs, such as USDA. Chapter One noted that universities find it easier to share costs when the federal agency mission is closely aligned with other university funders, such as state governments. The USDA has this characteristic. Of its $1 billion budget for research activities, most is for agricultural extension services, which require state or university matching. The federal government defines its share of the program as omitting facilities and administrative costs; these costs are paid by the matching funds. USDA's budget for competitive research project grants is just $84 million and is subject to congressional limitations on F&A costs. Universities appear to find funds to share the F&A costs on this $84 million.

Would it be advisable to extend these provisions to other agencies? It would be more difficult for universities to share F&A costs at the same level with NIH, whose total outlays to universities are $8 billion. In contrast to USDA, most of NIH's outlays to universities, $7 billion, goes for competitive research project grants. Biomedical research does not share the history of state funding for agriculture and is not therefore positioned to attract as much cost-sharing. Thus USDA is able to attract good proposals paying lower costs than NIH. USDA officials are concerned that limits on F&A costs may discourage some scientists from competing for grants. If the USDA were seeking to fund a larger competitive research grant program, this concern would be even more significant.

Even if other sources were tapped to create more supplements to the NIH $7 billion in research project grants, the social consequences could be undesirable. Universities might reduce support for other functions, such as education, or reduce investments in new and renovated facilities.

Although a fundamental change in the philosophy of reimbursing F&A costs is not under consideration, several options for further moderating F&A rates have been recently considered, and some of them have been implemented. Most special studies have now been eliminated. Special studies for libraries have been allowed, but the data show that overall negotiated rates for libraries have remained modest and almost constant. OMB considered benchmarking to manage the costs of new facilities that will be used in federal research. But because of the inherent difficulties in developing standard cost templates for research facilities that span a wide variety of needs, OMB selected a different approach. Circular A-21 now requires that major facilities be subject to a review of reasonableness, including comparison of costs with relevant construction data and a review as part of the Circular A-133 audit.

With the cap on administration in place, questions may be raised about controls applied to the facilities portion of F&A rates. The trend data indicate that there has been about a two percentage point increase in negotiated rates for facilities over the past decade. Negotiated rates for operations and maintenance have actually decreased, while infrastructure rates have increased. It appears that the increases in infrastructure rates stem from modernization of research buildings through renovation and new construction. Some of this modernization may have been encouraged by the allowance of interest expenses for research facilities, beginning in 1982. The survey data cited in Chapter Five indicate that universities have increased research facility space by 28 percent and substantially upgraded the average capability of their research space.

Ultimately, how much the federal government should provide for these infrastructure costs is a question of policy. But if the federal government reduces support for infrastructure, universities may well opt not to construct new facilities or modernize old ones. In this case, universities will have less capacity to pursue scientific research.

There may be other options for reducing federal outlays for F&A costs. If universities could reduce their costs, the government would be able to spend less without adverse effects on other programs. Chapter Six reviewed a broad spectrum of laws and regulations that contribute to both facilities and administrative costs. Streamlining the requirements imposed by law and regulation would enable universities to lower their costs and the federal government to reduce outlays for facilities and administrative costs.

Chapter Eight

ISSUE 6: OPTIONS FOR CREATING AN F&A DATABASE

A database of federal research F&A costs could be created and maintained. Doing so would require an organization within the government to take responsibility for operating it. Furthermore, adequate funding would be needed to ensure that this operation was properly staffed to design, maintain, and keep accurate F&A rate data.

Existing data systems capture some information already. Currently, ONR and DHHS break down negotiated F&A rates into major cost components. Formalizing the existing data collection within ONR and DHHS and standardizing between the two agencies could enable ongoing monitoring of the types of calculations in this report.

There are opportunities to expand some aspects of the data collection in ways that could support policy analysis. OMB is now developing a new standard format for universities to submit their F&A proposals to the cognizant agencies. If implemented, this format will offer an improved way of capturing certain data that supplement the rate and component information in university proposals. Information that may be covered on the standard format includes the estimated MTDC base for the next fiscal year and the number of square feet of space allocated to research, instruction, and other functions.

A particularly useful figure to collect through this new system would be the amount of estimated MTDC base eligible for full F&A recovery as opposed to amounts subject to a cap on F&A reimbursement. Knowing this amount—at least for large universities—would allow the government and researchers to estimate the effects of changes in F&A reimbursement. Typically, changes in F&A reimbursement will have no effect on grants already subject to a cap on reimbursement. Changing a rate from 50 to 48 percent will not affect grants with F&A capped at a smaller percentage. Changes affect only grants and contracts with full F&A reimbursement. An estimate of that quantity would be helpful in computing the impact of proposed changes.

The standard format, however, applies specifically to *proposals* rather than negotiated rates and components. Although proposal information alone would be of some value, it would be more useful to have the standard format items updated during the negotiation process to reflect the actual negotiated rates and their underlying components. Updating the proposal information, however, is not always straightforward. Negotiators may take a "bottom-line" approach by agreeing on a rate with a university without any specific accord on the assumptions and computations used to calculate it. The government will have to make a tradeoff between generating more accurate data on rate components and assumptions and streamlining the negotiating process.

Information on actual government *expenditures* for F&A reimbursements would be far more complex to capture, because it is generated within individual programs throughout the government. But a modest effort could still generate useful information. As discussed in the introduction, NIH awards the majority of federal research funds to universities. Because NIH already compiles a report on F&A costs (for grants with no limits on F&A reimbursement), a good deal of this information is available now. Because NIH is not representative of every agency, there might be value in having other agencies compile similar information. USDA already produces information on awards at an aggregate level. We caution, though, that because a few agencies account for almost all federal research funding to universities, expanding these data mandates to all federal agencies appears to be a large burden for a small incremental value.

Chapter Nine

SUMMARY AND CONCLUSIONS

As the research partnership between the federal government and universities evolved, federal agencies developed principles for reimbursing both the direct costs of research and some of the costs of facilities and administration. The reimbursement of these costs has long been the subject of congressional interest. In 1998, Congress asked for an investigation of issues related to this topic. In conducting an analysis of these issues, we have been hampered, in some cases, because the government does not maintain convenient databases from which to extract the requested information. The accessible government data contain information on *negotiated* facilities and administrative rates. Our analysis of these data shows that these negotiated rates have remained about constant for a decade, but we lack data on actual federal outlays for F&A costs. The data we do have are consistent with the findings based on negotiated rates.

Because we have to rely on incomplete data for actual outlays by agencies and receipts by universities, we can only make approximations in these areas. On average, about 31 percent of total true costs appear to be for facilities and administration. The share of federal outlays that pays for F&A costs is between 24 and 28 percent. Based on the difference between these figures, we conclude that universities are sharing in facilities and administrative costs. Overall, we estimate that the federal government does not reimburse between $0.7 and $1.5 billion in facilities and administrative costs allocated to federal projects based on negotiated F&A rates. Our analysis indicates that the federal government pays between 70 and 90 percent of the total negotiated amount for F&A costs.

Because universities report a total level of support for research from their own funds of about $5 billion, it appears that these unreimbursed facilities and administrative costs represent about one-fifth of the university funds devoted to research. The remainder of the $5 billion amount funds expenses of two types. One type is the universities' sharing in the *direct costs* of some projects, in particular by subsidizing faculty time. The other type of expense is funding for complete research projects by universities. The universities are voluntary participants in this system. Universities and their faculty are interested in attracting federal research support and have been willing to share in the costs.

Although universities clearly exercise some discretion in deciding how to staff administrative offices and how to construct facilities, many of the costs of facilities and administration derive from requirements in federal, state, and local law. These laws and regulations support a number of objectives, including the desire to protect the health and safety of humans and animals and to promote good stewardship for federal research funding. But they impose real costs.

In terms of the reasonableness of F&A costs in universities, our direct evidence is limited. What evidence we have indicates that the underlying cost structures in universities have lower F&A costs than federal laboratories and industrial research laboratories do. Because of specific limitations on university F&A reimbursement, such as the administrative cap, the actual amount awarded to universities for F&A costs is likely to be even lower than the amount cost structure comparisons would indicate.

Despite concerns about rates, average F&A rates have held steady for a decade. As administrative rates have declined because of the imposition of the administrative cap, facilities rates have increased.

Facilities rates have increased partly because of a change in federal policy that allows the inclusion of interest costs on new construction to be included in rate negotiations. Universities appear to have undertaken modernization especially during the 1990s, increasing research space by 28 percent. Although F&A rates now include more for construction components, the operations and maintenance component of rates has declined, perhaps because newer facilities are more efficient.

Overall, the research partnership between the federal government and universities is widely praised for its contributions to the public welfare. In the context of the total relationship, facilities and administrative costs are a fraction of total costs, although they are very real costs to both universities and the federal government. Some steps can be taken to benefit both partners. Good fiscal stewardship in government and in higher education calls for both partners to agree on a set of rules for reimbursing these costs. Because universities are in a position of making investments in their faculty, other personnel, and facilities that are expected to last for decades, universities have a strong preference for stability and predictability in the rules for cost reimbursement.

If the federal government pressed for greater cost-sharing by universities, it might get more. However, these additional funds would have to come from somewhere. We do not know how universities would finance additional cost-sharing. Universities faced with reduced federal reimbursement for facilities and administration might follow several strategies. They could reduce other projects within the $5 billion they already provide for research and allocate more as cost-sharing for F&A costs. As an alternative, universities could slow investments in building new facilities or renovating old ones. Other possible sources of funds for greater cost-sharing on research could come from reducing internal funding for other missions, such as education, public service, or patient care. We lack data to indicate the choices that universities would make. It seems worthwhile to further investigate the options for universities to shift funding and the consequences of those shifts before contemplating major changes in reimbursement of F&A costs.

Appendix A

BRIEF HISTORY OF CIRCULAR A-21

Prior to the issuance of Circular A-21 by the Office of Management and Budget in 1958, each federal agency developed and maintained its own cost recovery policies. Earlier, in 1947, the Office of Naval Research negotiated the first set of principles to determine indirect cost rates; it was referred to as the "Blue Book," or *Explanation of Principles for Determination of Costs Under Government Research and Development Contracts with Educational Institutions*." The publication of the "Blue Book" acknowledged that universities were significantly different both organizationally and programmatically from commercial firms and required different cost principles to cover unique accounting practices (Knezo, 1995, p. 4, citing DHHS).

Circular A-21's 1958 issuance represented a concerted effort at the federal level to establish governmentwide cost principles by revising ONR's "Blue Book." Circular A-21 was issued to be "applicable to research and development grants, contracts, and other funding agreements between the federal government and educational institutions." (Knezo, 1995, p. 6.)

Since 1958, Circular A-21 has undergone numerous revisions. Below we summarize the significant revisions, by year of implementation. There are inherent difficulties in compiling an accurate and comprehensive roster of changes covering such a lengthy period. We relied on several secondary sources for many of the earlier changes: Knezo (1995 and 1999) and AAU (1988). Where possible, we cross-checked sources and referred to the text of the *Federal Register* announcements of Circular A-21 revisions. For the most recent revisions in 1996, 1997, and 1998), we relied directly on the text of the *Federal Register* announcements as cited in the bibliography (OMB, 1996, 1997, 1998).

1958: The original Circular A-21 was issued, applying to research and development grants and contracts between the federal government and educational institutions. A-21 defined direct and indirect costs, and it set standards for accountability, documentation, and consistency.

Institutions receiving less than $250,000 in awards were permitted to use a simplified method (short form) to calculate and allocate indirect costs.

1961 and 1962: Revisions of A-21 clarified and refined methods used in identifying, classifying, and distributing indirect costs.

1967 and 1968: Changes in A-21 involved modification of effort-reporting requirements.

1969: The federal funding limit was raised to $1 million for universities that wished to use the simplified method (short form).

Principles and guidelines to be used in determining costs for training and educational service agreements were established.

1973: The administration of the Circular A-21 was transferred to the General Services Administration from the OMB. Circular A-21's name changed to federal Management Circular 73-8.

1976: Standards for allowable costs were made more precise.

1979: OMB resumed administration of Circular 73-8 and the title reverted back to Circular A-21.

Modified total direct costs were established as the basis for calculating the distributions of indirect costs among projects.

The threshold was raised to $3 million in direct costs for institutions wishing to use the simplified method (short form).

1982: Revisions eased effort-reporting requirements to cover only work funded by the federal government, rather than all research, teaching, and administration. Effort reports were now allowed to be filled out by persons other than the researchers.

The interest costs of debt directly associated with buildings and equipment supporting research were made allowable.

1986: Fixed allowance created for departmental administration (specifically of academic department heads, faculty, and other professional staff) that could be charged to research at 3.6 percent of MTDC. (The allowance was first set at a rate of 3.0 percent in June 1986 and revised to 3.6 percent in December 1986.)

1991: Costs in the administrative category were subject to a 26 percent cap. Ambiguities in interpretation of the circular to prevent shifting capped indirect costs to uncapped costs were removed.

Some costs were excluded, such as alcoholic beverages; entertainment; alumni activities; housing and personal expenses of officers; defense and prosecution of criminal and civil proceedings, claims, appeals, and patent infringements; and trustees' travel.

Assurances were required from universities that reimbursement for buildings would be used exclusively for research facility expenditures.

1993: Seven categories of cost categories (pools) were aggregated into two general categories: facilities and administration.

An option was created for schools to claim an allowance of 24 percent of MTDC for the administrative portion of indirect costs, or a percentage equal to 95 percent of the most recently negotiated rate for administrative cost pools, whichever is less. If schools elected to use the lower cap, they would not be required to prepare the paperwork necessary to document rates.

The time that predetermined fixed indirect costs rates could be used was extended from three to four years.

Government cost accounting standards and required disclosure of cost accounting practices were imposed.

The threshold in direct costs for institutions wishing to use the simplified method (short form) was raised to $10 million.

1996: "Facilities and administrative costs" replaced the phrase "indirect costs."

Four cost accounting standards applicable to educational institutions were incorporated into A-21.

Institutions receiving more than $25 million in federal sponsored agreements subject to A-21 were required to disclose their cost accounting practices by the submission of a DS-2 Disclosure Statement prescribed by the Cost Accounting Standards Board.

Use allowance and depreciation methodologies were clarified.

The threshold for capitalizing equipment was raised from $500 to $5,000 (or the institution's own selected capitalization threshold for its own accounting records, if lower).

Employee dependent tuition benefits were disallowed in fringe benefit calculations.

Interest costs on capital assets of more than $500,000 must be supported by an analysis of lease versus purchase. The interest costs of the cheaper alternative are allowable.

The F&A rates in effect at the start of a sponsored agreement must be used over the life of that agreement. Award levels for future years of a single agreement may not be adjusted based on changes in F&A rates.

The negotiation responsibilities for cognizant agencies were outlined.

Proposed elimination of special cost studies to allocate utility, library, and student services costs, effective 1998. (This provision was amended in 1998.)

1997: Conditional exemptions from OMB's grants management requirements were established for certain federal grants programs with statutorily authorized consolidated planning (for certain state-administered nonentitlement grants programs).

1998: Revisions were made establishing review and documentation requirements to ensure the reasonableness of the costs of large research facilities.

Utility costs adjustment recovery: 1.3 percent rate adjustment in lieu of special cost studies for 66 named universities, with review in 2002.

Elimination of special studies to determine library cost deferred.

Guidance on the calculation of depreciation and use allowances on buildings and equipment added.

Trustees' travel expenses were allowable.

Allowed universities that use the simplified method (short form) to use either salaries and wages or MTDC as a base to distribute their facilities and administrative costs.

Appendix B

RATE TYPES ALLOWED IN CIRCULAR A-21

Circular A-21 lists four types of rates:

Predetermined rates

Negotiated fixed rates and carry-forward provisions

Provisional and final rates

Negotiated lump sum.

The following paragraphs describe each type of rate, using extracts from Circular A-21 in quotations and additional explanations.

Predetermined rates for F&A costs. "Negotiation of predetermined rates for F&A costs for a period of two to four years should be the norm in those situations where the cost experience and other pertinent facts available are deemed sufficient to enable the parties involved to reach an informed judgment as to the probable level of F&A costs during the ensuing accounting periods." The predetermined rate is based on accounting data from a base period, a recently closed fiscal year of the institution, with any adjustments for projected changes mutually agreed on between the institution and its cognizant agency for negotiation.

Negotiated fixed rates and carry-forward provisions. In the past, this was a common method of setting F&A recovery rates. A fixed rate was negotiated in advance for a fiscal year based on forecasts and compared with actual allowable costs after the year is over. "The over- or under-recovery for that year may be included as an adjustment to the F&A cost for the next rate negotiation."

Provisional and final rates for F&A costs. "Where the cognizant agency determines that cost experience and other pertinent facts do not justify the use of predetermined rates, or a fixed rate with a carry-forward, or if the parties cannot agree on an equitable rate, a provisional rate shall be established. To prevent substantial overpayment or underpayment, the provisional rate may be adjusted by the cognizant agency during the institution's fiscal year. Predetermined or fixed rates may replace provisional rates at any time prior to the close of the institution's fiscal year. If a provisional rate is not replaced by a predetermined or fixed rate prior to the end of the institution's fiscal year, a final rate will be established and upward or downward adjustments will be made based on the actual allowable costs incurred for the period involved."

Negotiated lump sum for F&A costs. "A negotiated fixed amount in lieu of F&A costs may be appropriate for self-contained, off-campus, or primarily subcontracted activities where the benefits derived from an institution's F&A services cannot be readily determined." This arrangement is generally not used. Rather, for small institutions, the short form is used.

BIBLIOGRAPHY

FEDERAL GOVERNMENT PUBLICATIONS

DHHS Working Group on the Costs of Research, "Management of Research Costs: Indirect Costs," November 1991.

GAO, "University Research: Effect of Indirect Cost Revisions and Options for Future Changes," GAO/RCED-95-74, 1995.

_____, "Regulatory Compliance for NIH Grantees," GAO/HEHS-96-90R, 1996.

_____, "Assuring Reasonableness of Rising Indirect Costs on NIH Research Grant—A Difficult Problem," Washington, D.C.: U.S. General Accounting Office, March 16, 1984.

_____, "Regulatory Burden: Some Agencies' Claims Regarding Lack of Rulemaking Discretion Have Merit," Washington, D.C.: U.S. General Accounting Office, GAO/GGD-99-20, 1999.

House Science Committee, report on H.R. 2282, July 11, 1991.

_____, report, April 21, 1997.

Knezo, Genevieve J., "Indirect Costs at Academic Institutions: Background and Controversy—Issue Brief," Washington, D.C.: The Congressional Research Service, updated February 3, 1999.

_____, "Indirect Costs for R&D at Higher Education Institutions: Annotated Chronology of Major Federal Policies," Washington, D.C.: The Congressional Research Service, updated January 24, 1995.

National Institutes of Health, DHHS, "Indirect Costs," February 1999.

National Science and Technology Council, "Renewing the Federal Government–University Research Partnership for the 21st Century," Washington, D.C., April 1999, accessed at http://www.whitehouse.gov/WH/EOP/OSTP/html/rand/index.htm.

National Science Foundation, "Federally Sponsored Research: How Indirect Costs Are Charged by Educational and Other Research Institutions," 1991.

_____, "Scientific and Engineering Research Facilities at Colleges and Universities 1998: An Overview," 1998.

_____, WebCASPAR data system, 1999, accessed at http://caspar.nsf.gov/

Office of Management and Budget, "Cost Principles for Educational Institutions," Notice: Action: Final Revision and Recompilation of OMB Circular A-21, in *Federal Register,* Vol. 61, No. 90, 1996, pp. 20879–20941.

_____, "Government-wide Grants Management Requirements," Action: Final Revision of Circulars A-21, A-87, A-102, A-110, and A-122 and Interim Final Revision of OMB Circular A-110, in *Federal Register,* Vol. 62, No. 168, 1997, pp. 45933–45936.

_____, "Cost Principles for Educational Institutions," Action: Final Revision and Interim Final Revision of OMB Circular A-21, "Cost Principles for Educational Institutions," in *Federal Register,* Vol. 63, No. 104, 1998, pp. 29785–29792.

Office of Science and Technology Policy, "Renewing the Promise: Research-Intensive Universities and the Nation," Washington, D.C., December, 1992.

_____, "A Renewed Partnership: An Examination of Federal Government–University-Industry Interactions in U.S. Research and Higher Education in Science and Engineering," 1986.

Senate Committee on Labor and Human Resources, report, October 15, 1997.

FEDERAL GOVERNMENT WEB SITES

Animal Welfare Act As Amended (7 U.S.C. 2131–2156), accessed at http://www.nal.usda.gov/awic/legislat/awa.htm.

Federal Register through the Government Printing Office, accessed at http://www.access.gpo.gov/su_docs/aces/aces140.html.

National Institutes of Health, Division of Animal Welfare, Office for Protection from Research Risks, Frequently Asked Questions About the Public Health Service Policy on Humane Care and Use of Laboratory Animals, accessed at http://www.nih.gov/grants/oprr/faq_labanimals1997.htm.

Office of Management and Budget, OMB Circulars, accessed at http://www.whitehouse.gov/OMB/circulars/index.html.

Public Health Service Policy on Humane Care and Use of Laboratory Animals, revised September 1998, reprinted March 1996, accessed at http://www.nih.gov:80/grants/oprr/phspol.htm.

NONGOVERNMENTAL ORGANIZATION PUBLICATIONS

Arthur Andersen, LLP, "The Costs of Research: Examining Patterns of Expenditures Across Research Sectors," prepared for the Government-University-Industry Research Roundtable, Chicago, Ill., March 1996.

Association of American Universities, "Indirect Costs Associated with Federal Support of Research on University Campuses: Some Suggestions for Change," Washington, D.C.: AAU, December 1988.

Council on Governmental Relations, "Indirect Cost Rates at Research Universities—What Accounts for the Differences," Washington, D.C.: COGR, November 1987.

_____, "Managing Externally Funded Programs at Colleges and Universities: A Guideline to Good Management Practices," Washington, D.C.: COGR, May 1998a, accessed at http://www.cogr.edu/good.htm.

_____, "Factors That Influence Facilities & Administrative Cost Rates at Research Intensive Universities," Washington, D.C.: COGR, June 1, 1998b, accessed at http://www.cogr.edu/finalvariances.htm.

Federal Coordinating Council for Science, Engineering, and Technology, "In the National Interest: The Federal Government and Research Intensive Universities," 1992.

Federation of American Societies for Experimental Biology, "Graduate Education: Consensus Conference Report," 1997, also 1998 edition.

National Research Council, "Interim Report: Approaches to Cost Recovery for Animal Research: Implications for Science, Animals, Research Competitiveness, and Regulatory Compliance," 1998, accessed at http://www4.nas.edu/cls/ilarhome.nsf/web/cost_recovery?OpenDocument.

Norris, Julie T., and Jane A. Youngers, "Sponsored Programs Offices in Higher Education: A Continuing Evolution Responding to Federal Requirements," Washington, D.C.: COGR, accessed at http://www.cogr.edu/Norris Youngers.htm.

Rosenzweig, Robert M., "The Politics of Indirect Costs," Washington, D.C.: COGR, August 1998, accessed at, http://www.cogr.edu/Rosenzweig.htm.

"Stresses on Research and Education at Colleges and Universities: Institutional and Sponsoring Agency Responses," report of a collaborative inquiry conducted jointly by the National Science Board and the Government-University-Industry Research Roundtable, August 1994.

"Stresses on Research and Education at Colleges and Universities: Phase II," transcript from a meeting at the National Academy of Sciences, February 26, 1997.

ARTICLES AND BOOKS

Andrews, Rebecca, "Hazardous Waste Disposal: An Offal Problem for Laboratories," *The Scientist,* Vol. 5, No. 1, January 7, 1991.

Bennett, B. T., M. J. Brown, and J. C. Schofield, *Essentials for Animal Research: A Primer for Research Personnel,* Second Edition, revised October 1994.

Fairweather, James, Steven, "Reputational Quality of Academic Programs: The Institutional Halo," *Research in Higher Education,* Vol. 28, No. 4, 1988, pp. 345–355.

Feller, Irwin, "Matching Fund and Cost-Sharing Experiences of U.S. Research Universities," prepared for the National Science Foundation, March 1977.

_____, "The Changing Academic Research Market," State College, Pa.: Pennsylvania State University, 1995.

_____, "Social Contracts and Institutional Support of Academic Research," State College, Pa.: Pennsylvania State University, 1996.

_____, and Roger Geiger, "The Dispersion of Academic Research in the 1980s," *The Journal of Higher Education,* Vol. 66, No. 3, 1995.

Goldman, Charles, letter from university official, June 10, 1999.

Grunig, Stephen D., "Research, Reputation, and Resources: The Effect of Research Activity on Perceptions of Undergraduate Education and Institutional Resource Acquisition, *Journal of Higher Education,* Vol. 68, January/February 1997, pp. 17–52.

Lakoff, Sanford A., "Accountability and the Research Universities," in Bruce L. R. Smith and Joseph J. Karlesky, eds., *The State of Academic Science: Background Papers,* Vol. 2, New York: Change Magazine Press, 1978, p. 173.

Massy, William F., and Jeffrey E. Olson, "Overhead Diversity: How Accounting Treatments, Facilities Economics, and Faculty Salary Offsets Affect University Indirect Cost Rates," Stanford, Calif.: Stanford Institute for Higher Education Research Discussion Paper, November 1991.

May, Robert M., and Stuart C. Sarson, "Revealing the Hidden Costs of Research," *Nature,* April 8, 1999.

McGuire, Joseph W., Marie L. Richman, and Robert F. Daly, "The Efficient Production of 'Reputation' by Prestige Research Universities in the United States," Journal of Higher Education, Vol. 59, July/August 1988, pp. 365–389.

Noll, Roger G., and William P. Rogerson, "Chapter Five: The Economics of University Indirect Cost Reimbursement in Federal Research Grants," in Roger G. Noll, ed., *Challenges to Research Universities,* Washington, D.C.: Brookings Institute, 1998.